Electrical Installation Practice
Book 1

Electrical Installation Practice
Book 1

Fifth edition

H. A. Miller

Revised by
R. D. Puckering
LCG, BEd
City of Westminster College

OXFORD
BLACKWELL SCIENTIFIC PUBLICATIONS
LONDON EDINBURGH BOSTON
MELBOURNE PARIS BERLIN VIENNA

© R. D. Puckering 1989, 1993

Blackwell Scientific Publications
Editorial Offices:
Osney Mead, Oxford OX2 0EL
25 John Street, London WC1N 2BL
23 Ainslie Place, Edinburgh EH3 6AJ
238 Main Street, Cambridge,
    Massachusetts 02142, USA
54 University Street, Carlton,
    Victoria 3053, Australia

Other Editorial Offices:
Librairie Arnette SA
2, rue Casimir-Delavigne
75006 Paris
France

Blackwell Wissenschafts-Verlag GmbH
Meinekestrasse 4
D-1000 Berlin 15
Germany

Blackwell MZV
Feldgasse 13
A-1238 Wien
Austria

Fourth edition published 1989
Fifth edition published 1993

Set by DP Photosetting, Aylesbury, Bucks
Printed and bound in Great Britain at
the University Press, Cambridge

DISTRIBUTORS

Marston Book Services Ltd
PO Box 87
Oxford OX2 0DT
(*Orders:* Tel: 0865 791155
        Fax: 0865 791927
        Telex: 837515)

USA
Blackwell Scientific Publications, Inc.
238 Main Street
Cambridge, MA 02142
(*Orders:* Tel: 800 759-6102
            617 876-7000)

Canada
Oxford University Press
70 Wynford Drive
Don Mills
Ontario M3C 1J9
(*Orders:* Tel: 416 441-2941)

Australia
Blackwell Scientific Publications Pty Ltd
54 University Street
Carlton, Victoria 3053
(*Orders:* Tel: 03 347-5552)

British Library
Cataloguing in Publication Data
A Catalogue record for this book is
available from the British Library

ISBN 0–632–03524–2

Library of Congress
Cataloging-in-Publication Data

Miller, Henry Arthur.
    Electrical installation practice /
    H.A. Miller. — 5th ed. / revised by
    R.D. Puckering.
        p.   cm.
    Includes index.
    ISBN 0-632-03524-2
    1. Electric wiring, Interior—Handbooks,
manuals, etc.   2. Electric apparatus and
appliances—Great Britain—Installation—
Handbooks, manuals, etc.   I. Puckering,
R.D.   II. Title.
TK3271.M49   1993b
621.319′24—dc20   93-7062
                                            CIP

Contents

Preface

This well known series of books on the craft practice aspects of electrical installation work has been updated to meet the requirements of the 16th Edition of the IEE Wiring Regulations (BS7671), together with the new emphasis on practical competence required for the City and Guilds 236 Course. The course has been developed in consultation with the industry's lead bodies and has the full seal of approval of the National Council for Vocational Qualifications (NCVQ), a body set up by the Government to establish a framework of qualifications that meet agreed national standards. Qualifications will be made up of a number of units which are recognisable and have value in a particular employment.

The first step in any practical work is to learn how to work safely and prevent accidents. This book explains the law and regulations, and gives practical examples learnt from long experience, together with check-lists of dos and don'ts, and what to do when accidents happen. The book covers installation practice, associated craft theory, safe working practice, and a study of the electrical industries. This new edition has been expanded considerably to include some of the electrical principles underpinning electrical installation practice.

Book 1 is part of a series of books covering the City and Guilds 236 Electrical Installation course. It introduces students to the electrical industry, and explains the installation of electrical supplies, the installation of wiring systems, lighting circuits, small power circuits and basic inspection and testing procedures. It takes you step-by-step through the basic practical skills needed in installation work and covers the first year of the course.

Book 2 takes you through the second half of the City and Guilds 236 Part I course. It goes more deeply into some of the work covered by Book 1 and then introduces you to new topics such as the planning of electrical installations, wiring systems such as cable tray and trunking, and the installation of cookers, water heaters, fluorescent lighting and space heating. The inspection and testing of installations to the new 16th edition of the IEE Regulations is included and the selection of cables and the factors determining their choice is fully discussed.

Book 3 takes you through the City and Guilds 236 Part II course. It goes more deeply into some of the work covered by Books 1 and 2 and then introduces you step-by-step to new topics covered by the Part II syllabus, such as the installation of discharge lighting, electric motors, fire alarms and security systems. It goes on to describe the different organisations which operate both in and alongside the electrical contracting industry, and the rules and conditions of employment under which you will work. Following chapters explain in a practical way the methods used to record progress on site by the use of bar charts, and the way that we use time-sheets, invoices, and material requisitions. Examples are given of the preparation of daywork sheets, with lots of hints and tips on how this should be done.

All the books have full explanations and clear illustrations to guide you through the correct procedures, and you can check your knowledge with the multi-choice questions at the end of each chapter.

R.D. Puckering

Acknowledgements

The Institution of Electrical Engineers
J. A. Crabtree & Co Ltd
M. K. Electric Ltd
Pyrotenax Ltd
National Power
PowerGen
Rawlplug Ltd
British Insulated Callander Cables Ltd
Wylex (Scholes)
Martindale
Chloride Gent

Chapter 1
The Generation and Transmission
of Electricity

1.1 The generation of electricity

The siting of power stations

The electricity supply for the vast majority of installations in Great Britain comes in the first instance from power stations run by Nuclear Electric, PowerGen, National Power, the National Grid Company or one of the Scottish Electricity companies. Good use is made of the natural resources to be found in this country, and the power stations are usually sited close to these resources. A lot of attention is paid to the aesthetic effect of these installations, especially on sites of great natural beauty, and where it is important, the companies make good use of landscaping, tree planting and local stone to blend these into the countryside.

Coal burning stations

Many of the power stations are still coal-fired for the purpose of raising steam to drive the turbo-generators which produce the electricity (see Fig. 1.1), so these power stations tend to be situated near the coal fields of north-east England, Yorkshire, Derbyshire, Wales and Kent in order to reduce the cost of transporation.

These stations produce large amounts of ash as a waste product, some of which is sold for use in road building or the manufacture of building blocks. Some of it is piped as a slurry of ash and water into settling lagoons, which are often old sand or gravel pits. After draining the water off, it is either left as an infill and the land returned to agricultural use, or it can be removed and used for other purposes.

Pollution is a problem, and the electricity companies who are leading authorities on this spend millions of pounds on the development, installation and operation of equipment to prevent dust, grit and sulphur from reaching the atmosphere.

Oil burning stations

There are not as many oil-fired power stations as there are coal-fired. Despite this fact, well over 13 million tonnes of fuel oil is burned every year. In order to reduce the expense of transporting the oil great distances, these stations are mainly constructed beside the oil refineries of south-west Wales, southern and south-east England. Others are near to the deep water ports, so that oil can be shipped in by

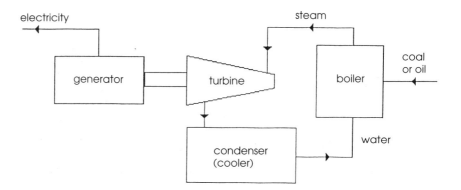

Fig. 1.1 The thermal process of generation.

tankers when it is in cheap and plentiful supply. With the discovery of North Sea oil, some of the coal-fired plants were converted to oil and the fuel piped in directly from the terminals. Oil-fired power stations are similar in many respects to the coal-fired ones, although there is no coal handling or pulverising plant.

Nuclear power stations

There are very few nuclear power stations in Great Britain, compared with such countries as France which generates most of her electrical energy this way. The amount of fuel used in this type of power station is very small, because in an advanced reactor, as much electricity can be made out of 1 kg of enriched uranium as from 50 tonnes of coal. The siting of these power stations is not easy, however, as great resistance to them is being put forward on environmental and ecological grounds. The used fuel, which is radioactive, is loaded into special thick steel flasks and taken by rail to British Nuclear Fuels Ltd for reprocessing.

Hydroelectric schemes

The three types of power station described above use either fossil fuels of nuclear fuel to heat water, and produce steam to drive the turbines. This is called the *thermal process*. It is not the only way of providing the prime mover (means of turning) to drive the generators. In areas where there is a good continuous flow of water, such as parts of Scotland and Wales, the natural force of the water can be used to drive the turbines. Environmentalists have mixed feelings about this type of power station. On the one hand, they are free from the pollution problems of the other types of station while, on the other hand, they flood large tracts of land to provide the reservoirs which are a requirement for this type of power generation. Once the initial cost of building the reservoirs, dams, penstock (pipelines) and generating plant is over, the running of these types of power station is relatively cheap. They do, however, depend on the weather to a certain extent to keep up a constant supply of water (see Fig. 1.2).

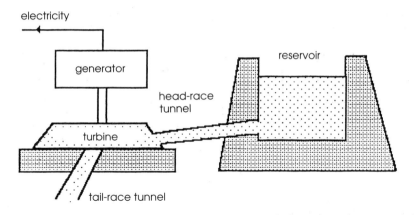

Fig. 1.2 The hydroelectric scheme.

Pumped storage schemes

A hydroelectric scheme which overcomes to some extent the problem of dependency on the weather is the pumped storage scheme. In simple terms, this consists of a hydroelectric power station situated between an upper reservoir and a lower one. The water is allowed to flow from the upper reservoir through the turbines and out into the lower reservoir, thus generating the electricity. When the water in the upper reservoir reaches a certain level, pumps using off-peak electricity take the water from the lower reservoir back up to the higher level. The level of the water is kept topped up by the natural rain-fall in the area. However, suitable sites are very difficult to find and are usually in places of great natural beauty. This apart, these power stations have much to offer; they are pollution-free, economical to run, and have the added advantage of being able to be brought on line at short notice. This is a particularly useful asset for the companies owning these types of station, for they can be brought into use when there are sudden high demands for energy, such as during an unexpected cold spell, for example.

Alternative means of generation

The electricity companies are constantly looking at alternative ways of generating electricity, and much pressure has been put on them recently by environmentalists to make use of what they see as pollution-free methods. It might be a good idea to look at some of these schemes and see just what they would entail.

Wind power   For this type of electrical generation, wind is required and lots of it. It has to be strong and fairly constant. In the British Isles, the Orkneys and Shetland Isles meet these requirements, and wind-driven generators are used to good effect in small communities there (see Fig. 1.3). To generate electricity on the scale required to equal the output of even a small, conventional type of power station, it is estimated that well over 100 square km of these would be required.

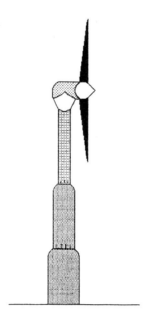

Fig. 1.3 Wind generator.

Wave power   Experiments are being carried out and schemes are being evaluated in the use of wave power for the generation of electricity. If the principle of converting the up-and-down motion of the waves into a reciprocating motion required to turn the generators is practical and economical, then Britain as an island would be indeed fortunate, as we are surrounded by the sea. At the time of writing, however, no positive schemes have emerged for the use of wave power.

Solar power   Solar panels, mounted on the roofs of houses, can prove very cost-effective when it comes to the heating of water, and in hot regions such as the Mediterranean, North Africa and the southernmost states of the United States of America, all the household requirements for heating and hot water can be obtained this way. Experiments are still going on, and scientists at Harwell Research Unit have found that an average of 0.2 kW of heat can be obtained per square metre an hour over most of Britain. At the moment, it does seem impracticable on a large scale, as some six million square metres of converting surface would be required to produce 2000 mW of electricity.

Tidal barrages   Of all the alternative schemes being evaluated at the moment, this seems to be the most viable. The principle behind the operation of this type of generating plant is to make use of the difference in water levels between high tide and low tide. Once again, Britain is fortunate in this respect as more than half the potentially suitable sites in Europe are situated here. The idea is to build a barrage across an estuary or bay, and allow sea water at high tide to flow in behind the barrier. When high tide is reached, gates are closed trapping the water behind the barrage. The water is allowed to flow out again at low tide, but first it must pass over the turbines of a generating station, thus producing electricity. This is a pollution-free method, but environmentalists will have to weigh this against the building of these stations at such beauty spots as the Bristol Channel and Morcambe Bay, two of the spots potentially suitable for this type of enterprise.

Gas   An increasing number of independent power producers are entering this market at a faster rate than was originally predicted, capitalising on the availability of gas from the North Sea and also the new combined cycle gas turbine (CCGT) technology. PowerGen hope to have their first CCGT on line in 1993. These stations help to reduce emission of carbon dioxide.

1.2 The transmission of electricity

Early developments in electrical distribution

In 1925, there were between 400 and 500 power stations in Great Britain. They were sited in or near the towns they were to serve and were, with few exceptions, independent of neighbouring power stations. Their size depended to some extent on the area they had to serve, but would on average be of around 5000 kW. There was no standardisation of voltage or frequency, in fact many were of the direct current (d.c.) type and not the alternating current (a.c.) that is used today. This independent policy not only created confusion when people moved equipment from one part of the country to another, but it hindered technical development too. Clearly something had to be done about the situation, and in 1926 an Act of Parliament was passed forming what was then known as the Central Electricity Board. The Board had the task of interconnecting the largest and most efficient power stations with a system of high voltage transmission lines operating at 132 000 V 50 Hz. This became known as the *grid system.*

The main purpose of the grid system was to operate the interconnected stations, so that the greatest number of units of electricity were generated by the then most efficient plant available. There were a number of other important advantages as follows:

- Standardisation of frequency and voltage;
- Less reserve plant required;
- Security of supplies;
- Energy transfer on a country-wide basis;
- Power stations sited near source of fuel;
- No need for stations to be in town centres.

As the demand for electricity increased, so did the size of the electricity supply industry to cope with it.

Later developments

In 1947, the whole of the industry was nationalised and in 1957, the name of the Central Electricity Board was changed to The Central Electricity Generating

Board (CEGB). The distances that the electricity was being carried had increased too and to achieve greater efficiency, the transmission voltages had to go higher and higher until we have what is known today as the *super grid*.

Recently the government privatised the electrical generating industry and it was split up into the separate companies mentioned at the start of the chapter.

The national grid

The large modern power stations generate electricity at something like 25 000 V, but, as we have seen, for efficient transmission over long distances, the voltage is increased. The reason for this is that for a given load the current is reduced with increased voltage; thus the cross-sectional are (*csa*) of the conductors can be reduced making a saving on cable costs. Energy losses are also reduced and transmission efficiency is improved, but this is offset to some extent by the need for higher towers (pylons) and better insulators. The voltages required to do this are 132 000, 275 000 and 400 000 V, and these are achieved by the use of *step-up* transformers. The electricity is transmitted at these high voltages to bulk supply points, where it is reduced by use of *step-down* transformers for distribution by the various regional electricity companies (RECs), at 33 000 V for heavy industry, 11 000 V for light industry and 415/240 V for commercial premises, farms and homes (see Information Sheet No. 1A).

1.3 The effects of the introduction of electricity

It would be hard to imagine the modern world as we know it today without the use of electricity; almost every facet of our lives is touched by this form of energy. Some of the ways it affects us are shown below:

In the home   It is used for lighting, heating and cooking, as well as operating such devices as vacuum cleaners, electric irons, kettles, and many other appliances.

Health and welfare   Our hospitals use it to good effect for operating theatre lighting, X-ray machines, sterilising equipment and a host of other devices.

Security   Fire alarm systems, intruder alarms and warning systems of various sorts are all operated by electricity for our safety and security.

Leisure pursuits   Imagine football and greyhound stadiums without flood-lighting, or track events without electronic timing and the photo-finish. These could not be done so effectively or as accurately without the use of electricity.

Entertainment   The home especially has been revolutionised by the use of such devices as television, video recorders, hi-fi equipment, etc. Theatres, cinemas, clubs, pubs and pop concerts have all been improved since the advent of this source of energy.

Information Sheet No. 1A Transmission and distribution of electricity.

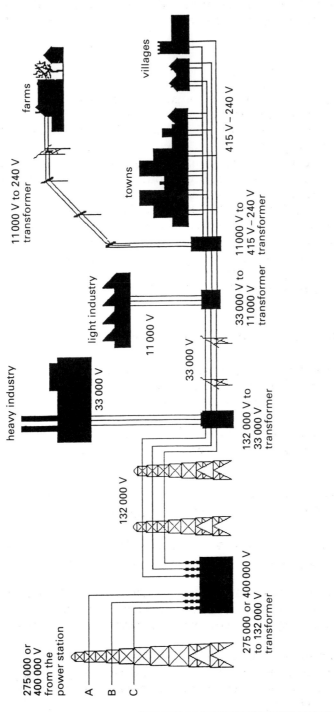

Labour saving devices   Devices such as electric drills, planers, saws and sanding machines have made life easier both at home and work, while in the home appliances such as microwave ovens, deep-fat fryers and food processors have taken some of the drudgery away from household tasks.

Communications   The modern marathon race is based on a Greek soldier running over 26 miles to deliver a message. Today, the same message could be delivered in seconds. Telephone systems, fax machines and satellite transmission for radio, television and communications have revolutionised the way we communicate.

Data storage   The familiar site of the general office of any big establishment is changing fast. Gone are the rows of filing cabinets, and typists slaving away over ancient typewriters or scribbling down letters in shorthand. This has been replaced by word processing stations and, instead of the filing cabinets, there are a few boxes of data storage disks..

The developments of the above, while spectacular in themselves, could not have been brought about without the use of electricity.

Test 1

Choose which of the four answers is the correct one.

(1) Power stations that burn fossil fuels use:

(a) Enriched uranium;
(b) The thermal process;
(c) Hydroelectricity;
(d) Pumped storage.

(2) The purpose of the *national grid* system is:

(a) To utilise available energy efficiently;
(b) To distribute low voltage supplies;
(c) To reach outlying farms;
(d) To save digging up the road.

(3) High voltages are used to transmit electricity because:

(a) They deter people from touching the conductors;
(b) Factories need the high voltages to work their machines;
(c) For a given load, the current is reduced with high voltages;
(d) Power stations generate electricity at these high voltages.

(4) The main advantage of a *pumped storage* scheme is:

(a) It uses up large amounts of rain water;
(b) It uses *off-peak* electricity;
(c) It provides fishing areas for anglers;
(d) It can *come on line* quickly.

(5) The *prime mover* for hydroelectric schemes is:

(a) Water;
(b) Steam;
(c) Gas;
(d) Nuclear fission.

Chapter 2
The Maintenance of Health and Safety at Work

2.1 Health and safety at work

Health and Safety at Work etc. Act 1974

Health and safety legislation originates from as far back as 1802, and since then has been added to by many new acts and regulations. Much of this legislation, however, is being brought together under the Health and Safety at Work etc. Act 1974. The Act seeks to replace the complex system of laws, rules and regulations with a less complicated arrangement based on Approved Codes of Practice.

The 1974 Act is designed to protect the health and safety of people at work, and places duties on employer and employee alike. It comes under the authority of the Health and Safety Commission, and the body responsible for putting the decisions of the Commission into practice is the Health and Safety Executive. Any person in breach of the provisions of the Act can, at the will of a visiting Inspector, be prosecuted, which, if successful, can result in a heavy fine or imprisonment. In fact, it is more usual for the Inspector to issue a prohibition notice stopping the activity or, if it is less serious, an improvement notice to remedy the fault in a given period of time. Prosecution under these circumstances would only follow if the notices were not complied with.

In practice, much of the safety aspects of doing particular jobs of work are taught as and when the operation is being carried out. For example, if you were drilling a metal bracket, you would be instructed that it would be advisable to wear safety goggles to prevent pieces of metal inadvertently entering the eye. These, and other safety aspects of particular jobs of work, will be discussed in future chapters; however, it might be a good idea to look at the more general implications of the 1974 Act as it affects people working in the electrical contracting industry.

2.2 The employer's responsibility

Under the Act, the employer is responsible as far as is practicable for the health, safety and welfare of all their employees at work. This applies particularly to the following:

Safe place of work

Employers are required to see that, as far as is practical, the place of work is safe for you to carry out your job. On building sites, the weather plays a big part in safety aspects, for example, slippery surfaces caused by ice, snow or water can present a hazard, especially at the entrances to sites and on concrete staircases open to the weather. This makes access to the place of work difficult and, therefore, in the interests of safety, these should be kept clear and treated with sand or salt to ensure safe passage for the employees. No one should be required to leap open trenches or climb mounds of loose earth, in order to get to their place of work. Designated safe routes should be provided for both vehicles and pedestrians, with plastic tapes or barriers indicating the extent of any danger areas. Staircases to upper floors should have temporary handrails provided and any openings in floors or edges to buildings should have safety barriers erected around them. Where people are working at different levels, there is always the chance that tools or building materials can fall to the lower floors. In situations such as this, toe boards should be provided on building edges and scaffolding (see Information Sheet No. 2A). In existing buildings, entrances and exits should be clearly identified and kept clear of obstructions. Gangways should be marked by yellow lines, and warning notices posted if used by vehicular traffic. The building should be adequately lit by artificial or natural light, and comply with local and national regulations regarding the building's structure, fire precautions, emergency lighting and other building services such as gas, electricity, telephones, etc.

Provision of safe plant and equipment

It is the employer's duty under the 1974 Act to see that all plant and equipment, whether it be fixed equipment in a workshop or portable equipment on site, is safe to use and in good working order. Regular checks should be carried out to see that equipment and plant complies with the appropriate standards with respect to the health and safety of employees (for electrical equipment see Chapter 3). We have seen that it is the employer's responsibility to provide safe access to the place of work and if this involves the use of access equipment (ladders, etc.) then these too should be maintained in a sound and serviceable condition (see Information Sheet No. 2B). It is the employee's duty to check the ladders, etc., before use, and to use them in the recommended way (see Information Sheets Nos 2C and 2D). Many items of equipment are now fitted with safety features such as fences on circular saws, perspex screens on grindstones, safety guards on rotating machinery and dust extraction equipment on cutting and forming machines. These should not be removed under any circumstances; they are fitted for your safety, and removal could make them dangerous for you or for people following after you. It is for this reason that employers are obliged under the terms of the Act to give their employees full instructions for the use of potentially dangerous equipment, and, where necessary, post working notices on the operation of systems and machinery. It is essential that these safety devices are regularly checked and are in place (see sections 2(1) and 2(2) of the Act).

Safe systems of work

Some of the jobs which people in the electrical contracting industry undertake are either potentially dangerous or are carried out in potentially dangerous situations.

Information Sheet No. 2A Scaffolding.

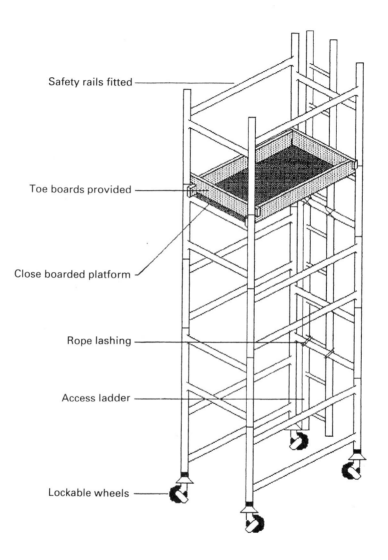

Safety rails fitted

Toe boards provided

Close boarded platform

Rope lashing

Access ladder

Lockable wheels

Information Sheet No. 2B Ladder safety check.

Check ladders for the following defects:

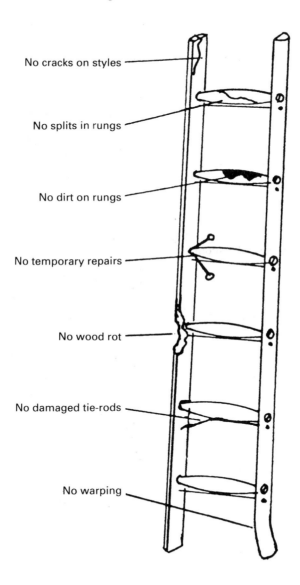

No cracks on styles

No splits in rungs

No dirt on rungs

No temporary repairs

No wood rot

No damaged tie-rods

No warping

Information Sheet No. 2C Erection of ladders and trestles.

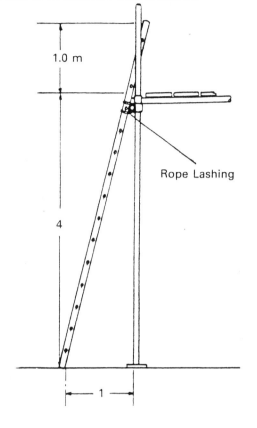

1.0 m

Rope Lashing

4

1

Rules for Ladders

(i) Ladder must extend 1.0 m above landing area
(ii) It must be firmly secured at the top
(iii) It must be at an angle of 75° (4 up 1 out)
(iv) It must be on a firm base.

Rules for Trestles

(i) The trestles must be on a firm base and both be level
(ii) The distance between the trestles must not exceed 1.3 m for 40 mm thick boards and 2.5 m for 50 mm thick boards
(iii) The total width of boards should exceed 450 mm
(iv) The boards should not be placed more than two-thirds of the way up the trestles and should not overhang by more than 4 times the thickness of the boards.

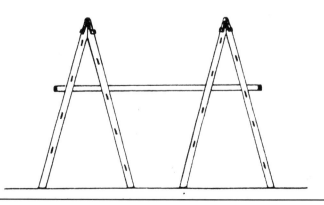

Information Sheet No. 2D Handling ladders.

1. With the ladder in the upright position rest it on the shoulder. If you are right-handed reach down with your right hand one rung lower than you would normally reach when standing erect. Lift by straightening the knees. Get your balance correct before moving off.

2. When moving ladders which are heavy or for long distances, then two people must be used as shown in the sketch.

It is for that reason that employers have to ensure that procedures and systems of work are designed to safeguard the health and safety of those putting them into practice. Potentially dangerous situations might be as follows:

- Close to high voltage equipment;
- Working in deep excavations;
- Petrochemical installations;
- Processes giving off toxic fumes;
- Nuclear plants.

Access to these areas is prohibited and before entry can be obtained a *permit to work* must be obtained. An example of such a permit is shown in Fig. 2.1. This is a working procedure, whereby people wishing to enter restricted areas must obtain permission to do so from an authorised person. This ensures the following safeguards are observed.

(1) The area is safe for you to enter;
(2) Other people are aware that you are in there;
(3) The task you are carrying out is permissible;
(4) It will be known when you have left the area.

Working systems such as this are for your safety and should under no circumstances be abused.

A number of safety systems installed for your protection, such as emergency stop buttons, residual current devices, fire alarm systems, etc., are not in constant use and therefore should be regularly tested at appropriate intervals to ensure that they are still operating correctly.

The workplace environment

This will depend to a large extent on the sort of contract upon which you are working. The conditions when working on existing industrial premises will to a large extent be fairly well established. The company or organisation responsible for the building will already have implemented a safety policy, which will be subjected to regular reviews by representatives of their management and workforce. The conditions will not therefore be under your employer's direct control, however, if certain adverse environmental conditions do exist, your employer should make you aware of these, and inform you of the precautions which you should take to avoid putting yourself and others at risk. If any safety equipment is required in order for you to carry out the work, this should be provided by your employer at no cost to yourself.

Examples of these types of environment are manufacturing processes, which use or produce toxic chemicals or fumes; grinding or cutting operations which produce fine particles of dust or fibres; manufacturing methods employing corrosive substances, and machines which produce a high pitched or excessively loud sound. Many of these processes can be dangerous to the eyes, lungs, skin or ears, and any goggles, masks, ear muffs, respirators or protective clothing, provided for your use, should be worn at all times whilst working in such areas.

DEPARTMENT OF THE CHIEF ELECTRICAL ENGINEER
PERMIT
PERMIT TO WORK
ON HIGH VOLTAGE ELECTRICAL APPARATUS OTHER THAN HIGH VOLTAGE CABLES

Serial numbers of associated Isolation Certificates ....

ISSUE

To: Name ....

  Grade .... Depot ....

I have read and
understand Parts 1
& 2 of the High
Voltage Safety Rules
and I hereby declare
that it is safe to
work on the
following high
voltage apparatus:

That the apparatus
is dead and isolated
from all live
conductors at the
following point(s):

Associated voltage and/or control transformer have also
been isolated.

The apparatus is
efficiently connected
to earth at the
following point(s):

1 ....
2 ....
3 ....
4 ....
5 ....
6 ....

Caution notices are posted at ....

Danger notices are posted at ....

Safety barriers are erected at ....

ALL OTHER PARTS ARE DANGEROUS

The following is the
work to be carried
out on the
apparatus:

Issued with the consent of .... SHIFT SUPPLY ENGINEER

SIGNED .... being an authorised person to issue a permit for the work.

Time .... Hours .... 19 .... Location ....

Fig. 2.1 Permit to work

The employees should make it their business to find out what working procedures or rules apply when working in these environments, as often *permits to work* are required before you can enter certain areas. Other potentially hazardous conditions are processes using inflammable materials; in this situation, all notices regarding fire prevention should be observed and the operatives should make themselves familiar with the recommended fire-drill. Further details on fire prevention can be found in Chapter 3.

On building sites where you are carrying out new work, the rules and procedures for safety may not be so well laid down as the above, although all the safety rules still apply. Adverse weather plays a big part in the sort of environmental conditions that exist, and although there is little an employer can do about this, wellington boots and water-proof clothing should be provided for employees spending large amounts of time outside. Not only does this mean that operatives are kept dry and warm, but it also means that the job can proceed even if the weather is a little inclement. Protective clothing of this nature is essential for people laying underground cables, erecting overhead lines and those installing outdoor equipment. Where people are working at different levels, there is always the chance that tools or building materials can fall to the lower floors; in these situations safety hats should be provided by the employer and worn by the operatives (see Fig. 2.2). Facilities on the bigger sites are run under what is known as *shared welfare schemes*; that is to say that all the employers of the various trades working on the site, have contributed, as part of the terms of the contract, to the supplying of certain welfare provisions. These often take the form of portable toilets, washrooms, canteens, first-aid facilities and drying rooms for wet outdoor clothes. Ask your employer what provisions have been made on the site you are working on.

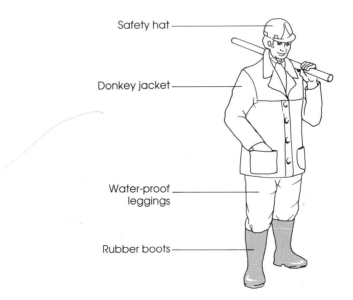

Safety hat

Donkey jacket

Water-proof leggings

Rubber boots

Fig. 2.2 Protective clothing

Information Sheet No. 2E Lifting objects.

1. Feet should be about 0.5 m apart, one foot slightly forward.
2. Knees should be bent slightly.
3. Back must be straight.
4. Arms must be kept close to the body.
5. Head should be erect with the chin tucked in.

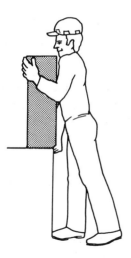

6. Take a firm grip of the object.
7. Lift by straightening the legs, not by pulling with the arms.
8. Keep the object close to the body and look in a forward direction.
9. Place the object on a firm base before letting it go.

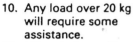

10. Any load over 20 kg will require some assistance.
11. Lift as before with one person giving the instructions.

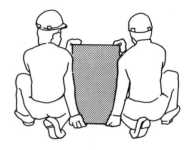

The handling, transportation and storage of equipment

The handling of equipment starts with the basic act of lifting a piece of equipment up and placing it in another position, for example, on to a bench or into the back of a van. It sounds so simple, yet the resulting back strain, due to incorrect posture when lifting or handling goods or equipment, is one of the major reasons for lost working days in industry. Information Sheet No. 2E shows the correct method of manually lifting loads. Any load over 20 kg in weight would normally need some form of assistance, and Information Sheet No. 2E shows methods of moving equipment by *team lifting*. The moving of really heavy equipment, such as large oil filled transformers or emergency generating equipment, is best left to experts trained in the moving of heavy plant. Not only are they well practised in the techniques of lifting and moving, but they will do it more quickly, more safely and without damage to the equipment. Advice should be sought when moving potentially dangerous substances, as incorrect procedures can result in serious spillage or even explosions. There are occasions, however, when the operative is required to use lifting gear, for example, off-loading equipment that has been delivered to site. This often takes the form of simple block and tackle, which is either manually operated or is powered by an electric motor (see Fig. 2.3). The use of this type of equipment is not beyond the scope of the average electrician, but it must be stressed once again that full instructions as to its proper use should have been received prior to attempting a lifting operation.

The moving of goods and equipment over long distances requires the use of some form of wheeled transportation. This, in its elementary form, could be a *sack barrow* or hand truck; these are the simple two-wheeled devices once much used

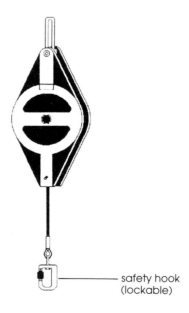

safety hook
(lockable)

Fig. 2.3 Block and tackle

by British Rail porters. Great care must be taken that these are not overloaded, and that you have a clear view of where you are going. Another method is to use a flat trailer. A great deal of equipment can be moved in this way, but care must be taken to see that it is securely fastened down and, if the equipment is circular in shape, to see that it is chocked so that it cannot roll (see Fig. 2.4). Some of these flat trailers have only three wheels, and care should be taken when turning corners with this design, as they can topple over on tight bends. Fork-lift trucks are, without doubt, most versatile pieces of equipment, not only for off-loading equipment, but for moving equipment too. Many of the goods sent to site are now dispatched on pallets; that is to say special timber frames designed so that the fork-lift truck can engage and disengage its forks without disturbing the load. This has improved the distribution of goods to such an extent that few premises or sites are without them. There are many restrictions on the use of fork-lift trucks regarding types of loads, heights they can be lifted, etc., and under no circumstances should petrol or diesel operated trucks be used inside buildings. Before attempting to make use of a fork-lift truck, make sure you are fully instructed in its use and that you are covered by insurance.

Storage of equipment on sites can be reduced to a minimum by having materials scheduled for delivery just prior to the time that they are wanted. Delays and alterations do occur and some storage facilities will have to be provided for this and the usual day-to-day items. Storage raises several important points which should be observed; the main ones are as follows:

- Expensive equipment and that on long delivery should be placed under lock and key;
- Equipment which is easily damaged should be protected from breakage by packaging, etc.;
- Certain items, such as galvanised cable tray, conduit and trunking, can be placed in outside compounds;
- Equipment which can be damaged by the weather should be placed under cover;

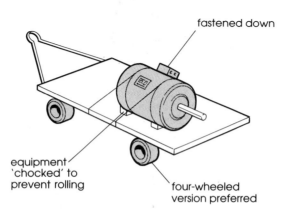

fastened down

equipment 'chocked' to prevent rolling

four-wheeled version preferred

Fig. 2.4 Flat trailer

- Goods should not be placed in designated gangways or walkways, as this can cause obstruction in the case of an emergency;
- Entrances and exits should be kept clear of goods and equipment for the same reason;
- Dangerous substances should be placed in designated compounds and be clearly marked;
- Inflammable substances should likewise be placed in approved compounds and be clearly marked;
- Locked storage facilities should be provided for employees, tools and equipment.

The training and instruction of employees

The training and instruction of the operatives in the electrical contracting industry is the responsibility of the employers in the industry. Not only are they responsible for training and instruction in necessary skills of the trade, but they are also responsible for seeing that operatives receive instruction on all aspects of safety, as outlined in the 1974 Act. Many of the employers in the electrical contracting industry have their apprentices trained under the Joint Industry Board (JIB) training scheme. Under this scheme, junior apprentices will be expected to take the City and Guilds 236 Craft Certificate (Part 1) in Electrical Installation and also pass a practical test called Achievement Measurement 1 (AM1). If successful, the apprentice will be graded Senior Apprentice (Stage 1). Following this, the City and Guilds 236 Craft Certificate (Part 2) in Electrical Installation is taken and, if successful, the apprentice is graded Senior Apprentice (Stage 2). When a further degree of skill and knowledge has been achieved, the apprentice will take a further practical test called the Achievement Measurement 2 (AM2). Success in this will complete the training and lead to eligibility for grading to Electrician.

The employer's responsibility does not end there, however, because with the increased use of labour-saving devices, many of them electrically operated, it is essential that operatives receive instruction in the use of these if accidents are to be avoided. Some of the more common devices needing specific instruction would be:

- Cartridge-fired fixing tools;
- Angle grinders and abrasive wheels;
- Tools using compressed air;
- Threading machines (electrical);
- Equipment using bottled gas;
- Certain testing equipment.

Operatives must also be fully instructed in working systems (see above) that are in operation in the area in which they are working, and advised of their responsibilities under the Health and Safety at Work etc. Act 1974. Every new entrant to the JIB training scheme is issued with a 'Code of Good Practice, Health and Safety at Work for the Electrical Contracting Industry'.

Safety policy

Safety policy in a small company may well be by word of mouth; bigger companies will find it almost impossible to see every one of their employees regularly. It is under these circumstances that the recommendations of the Safety Representative, and Safety Committee Regulations 1977, might prove to be useful. Safety policy, subject to regular review jointly by employers' and employees' representatives, is by far the best and easiest way of producing a common policy acceptable to all concerned. The safety policy should concern itself with all aspects of safety applicable to the type of work the company carries out, and should include things like the reporting of accidents, the keeping of accident registers, compliance with the Reporting of Injuries, Diseases and Dangerous Occurrences Regulations 1985 (RIDDOR) and instruction and training in safety matters.

Public liability

If work involving major alterations or additions to domestic or commercial premises takes place in a vacated building or an area that has been sealed off from the general public, then the environmental conditions are much like those on the building site above. Electrical contractors, however, are often asked to carry out work in buildings where people are still in residence or which are open to the general public. The 1974 Act makes the employer responsible for seeing that any work being carried out is done in such a way as not to endanger other persons who may not be in their employ, and this includes residents and customers as well as other tradespeople. Companies have by law to take out public liability insurance if their work can endanger people other than their own employees (employees are covered separately). Employers should make it quite clear to their employees how they should conduct the work under these circumstances, and they should supply any special requirements such as dust sheets, temporary lighting, etc., to protect the property and ensure the health, safety and welfare of all concerned.

2.3 Employees' responsibility

Employees too have responsibilities under the Act and, in general, these are as follows:

Responsibility to themselves and others

Employers can provide you with safety equipment and instruct you in the use of it. They can provide you with equipment to help you do your job more easily and efficiently, and see that this is kept in good condition, but they cannot cater for what is often called 'human error'. Most accidents don't just happen; they are caused by people's actions, or failure to take the appropriate action when faced with potentially dangerous situations. The person who knowingly uses a damaged piece of equipment is guilty of negligence, but so is the person who sees the damage and does not report it. You are responsible under the terms of the 1974 Act to see that your actions not only leave you free from hazard, but that they leave other people around you free from hazard too. We have seen how dangerous ice and snow can make surfaces. These dangers are not only attributable to weather, however; substances such as oil, grease, cutting compounds, paints and solvents spilled on the floor and not dealt with immediately can cause hazards. Even things like food, leaves from nearby trees and the off-cuts of conduit and pipe are dangerous if left underfoot. Clearly employees have some responsibility both to themselves and others to keep the workplace hazard free.

Follow the safety procedures

Employees must co-operate with their employers to carry out the duties of the Act, i.e., they must follow the safety codes laid down and make full use of any safety equipment provided. For example, any safety procedures such as the permit to work must not be abused by trying to take short cuts with paper work, etc. To do so would mean that the authorised person would have no way of knowing if you were in the area, and may take action that would place you in danger. When you are supplied with safety equipment, it might feel uncomfortable and you may not always see the need for it. A safety helmet, for example, feels very awkward at first, but if you persevere you will soon get used to it. By taking the helmet off and leaving it in the canteen or elsewhere, you are leaving yourself open to injury from falling objects, and if you were to be seriously injured it is unlikely that you would receive any sort of compensation if the employer could prove that you had been issued with a safety helmet.

Interference with articles or substances

Under the Act, no person may intentionally misuse or recklessly interfere with any article or substance, so as to put at risk the health and safety of themselves or others. Do not, for example, remove ladders from one part of the site to another without first checking that they are not in use. You may have someone stranded up a scaffold, or, worse still, they may try to come down a ladder that is no longer there. The construction industry uses many different substances that are potentially dangerous. Do not get into the habit of opening any tins or containers just to see what is inside; they may contain substances dangerous to the skin, or give off fumes which can seriously affect the eyes or throat. See *Electrical Installation Practice Book 3* (in this series) for further information on the Control of Substances Hazardous to Health (COSHH) Regulations.

Misuse of equipment

Employees must not misuse or interfere in any way with items of equipment provided for their health, safety or welfare. We saw earlier that fences, screens and guards are fitted to rotating machinery; these should not be removed under any circumstances, even if you feel that they are slowing the operation down. These are fitted for your safety and removal could mean at best that you were liable for prosecution under the terms of the Act, or at worst serious injury to yourself or others for which there would be little, if any, compensation.

Reasons for accidents

The construction industry has a worse record for accidents than any other single industry. We have seen that many of these accidents are due to human or environmental reasons; let's have a look at some of these:

Human   This can be due to a number of reasons, such as carelessness, tiredness, improper behaviour (horseplay) and dress, the taking of drugs for health reasons, drug abuse, the drinking of alcohol, poor eyesight, colour blindness, lack of experience, poor supervision, and lack of proper instruction.

Environmental   Taking the meaning of the word in its broadest sense, hazards could result from inadequate ventilation, poor lighting, hostile environment, overcrowding, inclement weather, unsafe plant and equipment, lack of safety equipment, or a dirty or untidy workplace.

Reducing the risk of accidents

The best way to reduce the risks of accidents is to try and remove the cause. Providing adequate lighting at the workplace would reduce accidents caused by people not being able to see things clearly, and also extend the working day in winter time. Locking dangerous substances in approved compounds so that people would not be tempted to interfere with them, causing danger to themselves and others, would remove another potential hazard.

If a hazard cannot be removed totally, then it might be possible to replace the risk with something less dangerous. For example, electric hand tools on site should be designed to be used on 110 V a.c. (55 V to earth) and not 240 a.c., thus reducing the risk of severe electric shock (see Chapter 3).

Hazards that cannot be removed or reduced can be guarded against; guards and fences can be put on machines, for example, or safety helmets or goggles issued, or hair nets provided for long hair, and other protective clothing supplied for your personal protection. Personal hygiene is important and barrier creams should be available for the hands, as well as the usual toilet and washing facilities.

It is necessary to develop a positive personal attitude towards safety in people's minds if the safety record is to be improved, and one of the best ways to achieve this is by safety instruction and education, and by a strong publicity drive pointing out the dangers. It is said that safety is expensive, but so too is the alternative, in lost time, damaged equipment, compensation for injuries and the needless loss of life.

Test 2

Choose which of the four answers is the correct one.

(1) The Act of Parliament concerning health and safety at work is:

(a) The Health and Safety Act 1974;
(b) The Health and Safety etc. at Work Act 1974;
(c) The Health and Safety at work etc. Act 1974;
(d) The Health and Safety at Work Act 1974.

(2) The above Act of Parliament is for the guidance of:

(a) Employees only;
(b) Members of the public only;
(c) Employers only;
(d) Employers and employees.

(3) An example of access equipment would be:

(a) A spare hacksaw;
(b) An extension ladder;
(c) A banker's card;
(d) Too much conduit.

(4) When lifting equipment the main muscles used should be the:

(a) Back muscles;
(b) Arm muscles;
(c) Muscles in the legs;
(d) Muscles in the hands.

(5) A safety precaution when using an electric drill would be to:

(a) Use a slow speed;
(b) Drill a little deeper than required;
(c) Stand on a pair of step-ladders;
(d) Wear protective goggles.

Chapter 3
The Need for Wiring Regulations

3.1 Electrical safety

The need for the IEE Wiring Regulations

As the use of electricity became more popular, it soon became clear that some unified form of regulations, concerning its safe installation into buildings, would be necessary if serious accidents were to be avoided. It was for this reason that the first edition of the *Rules and Regulations for the Prevention of Fire Risks Arising from Electric Lighting* was published in 1882. The early Regulations set out to achieve the following:

(1) To safeguard the users of electrical energy from shock;
(2) To minimise fire risk;
(3) To ensure as far as possible the safe and satisfactory operation of apparatus.

The use of electricity as a form of energy has expanded at a tremendous rate since those days, and new editions of the Regulations have been brought out to keep pace with these developments. The latest edition published by the Institution of Electrical Engineers (IEE) is the 16th edition and is entitled *IEE Wiring Regulations – Regulations for Electrical Installations*. The full scope and objectives of these is described briefly in the Regulations in Chapters 11 and 12.

Statutory requirements ✗

It is important to note that the IEE Wiring Regulations are not statutory; that is to say they are not law but only BS7671. However, several Acts of Parliament, notably the Electricity Supply Regulations 1988, and the Electricity at Work Regulations 1989, are in part commensurate with the IEE Wiring Regulations, and as these Acts are mandatory and have the force of law, it would be very difficult indeed to have an electrical installation connected to the public supply if it did not in fact comply with the IEE Wiring Regulations. No matter how complicated electrical installations have become, and how much the regulations have changed to accommodate them, the three basic objectives listed above are still as relevant today as they were then, and therefore it would be useful to look at them more closely.

3.2 Electric shock

How the human body can become part of an electric circuit

The *Collins English Dictionary* defines electric shock as follows:

'The physiological reaction characterised by pain and muscular spasm to the passage of an electric current through the body. It can affect the respiratory system and heart rhythm'.

How does this condition come about? It can come about in two ways:

(1) If we connect the two leads of an approved test lamp across the phase and neutral connections of a circuit as shown in Fig. 3.1, and the circuit is switched on, the current will flow through the lamp and the lamp's filament will glow. This is because the lamp has completed a circuit between phase and neutral. Now if a person places themselves across the phase and neutral connections of a circuit, as shown on Information Sheet No. 3A, they too will have completed a circuit and current will attempt to flow through them as shown, and they will receive an electric shock.

(2) Most people will appreciate that an electric shock can also be received by the touching of a current-carrying conductor or a live part of electrical equipment. How then does this shock come about? We have seen from the first example that for current to flow, a circuit of some kind must be completed. In the second example, however, we are only touching the current-carrying conductor and not the netural conductor as well; how then is the circuit completed? The circuit is completed, as shown on Information Sheet No. 3A, by the current passing once again through the body of the person, but this time it travels from phase to earth as shown. It will be seen that the current passes through the earth and returns to the star point of the supply authorities' transformer, which is in effect neutral, and so completes the circuit.

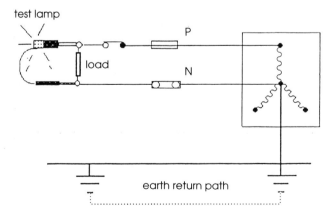

Fig. 3.1 Test lamp across the circuit

Information Sheet No. 3A Electric shock.

(i) Between poles

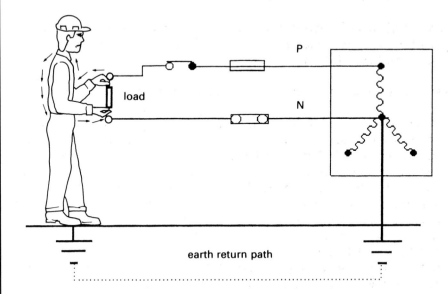

(ii) Phase to earth

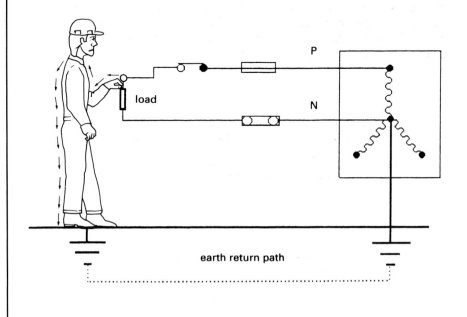

Action to be taken in the event of an electric shock

If a friend or colleague receives an electric shock and is still in contact with the source of supply you must act speedily to:

- Find the isolator for the circuit quickly, and turn off the supply; or
- Break the person's contact with the supply by the use of some insulating material such as a coat or a broom handle; and
- Take care not to come into direct contact with the person while they are touching the supply or you too will get a shock; and
- Lower the casualty to the floor taking care not to damage the head.

If the casualty is conscious, make him comfortable and give him reassurance. If you are in doubt about his condition, then get in touch with a doctor, and report the accident to the appropriate personnel.

Should the casualty be unconscious but breathing, loosen the clothing around his neck and waist and place the casualty in the recovery position (see Fig. 3.2). Keep a constant check on his breathing and pulse; improvise a suitable method to keep him warm and arrange for medical help if necessary.

When a casualty is found unconscious, but **not** breathing, then **take immediate action** and apply emergency resuscitation. There are two resuscitation techniques that you should be aware of:

(1) Mouth-to-mouth or mouth-to-nose;
(2) Holger–Neilson method.

Mouth-to-mouth resuscitation is by far the most commonly used form of resuscitation and is most effective in the event of electric shock. However, if the face has sustained injury, it may be more practical to use the Holger–Neilson method; both methods are shown on Information Sheets Nos 3B and 3C.

Isolation of electrical supplies

If electric shock is to be avoided, then it is highly desirable that the installation or equipment that is being worked upon is isolated from the mains supply. The IEE Wiring Regulations ask that, except for certain cases mentioned in Chapter 46 of the Regulations, every circuit shall be provided with means of isolation from the live supply conductors. In addition to the normal means of isolation at the mains

Fig. 3.2 The recovery position

Information Sheet No. 3B Mouth to mouth resuscitation.

1. Lay the casualty on his or
   her back and check
   the mouth for blockages.
   If possible raise the
   casualty's shoulders
   with padding of some
   sort.

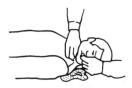

2. Make sure the head is
   well back and the
   air-way is clear.

3. Pinch the casualty's nose.
   Take a deep breath and
   seal the lips around
   the open mouth of
   the casualty.

4. Blow gently and firmly
   into the mouth; the
   chest should rise
   slightly as the lungs
   fill with your air. Repeat
   until casualty shows signs
   of recovery.

Information Sheet No. 3C Holger–Neilson method of resuscitation.

1. Place the casualty face downwards with the head to one side; check that the mouth is clear. Kneel in front of the casualty as shown and place both hands flat on the upper part of the back. Rock forward applying pressure with the hands.

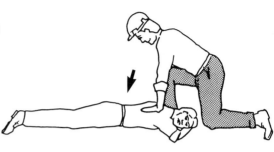

2. Rock backwards sliding the hands along and under the arm-pits. Grasp the upper arms as shown and lift. This will bring air into the lungs.

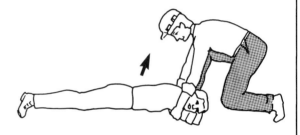

3. Lower the casualty gently down again.

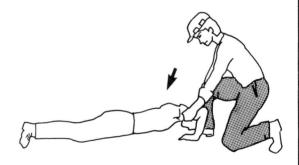

4. Repeat the operation over and over again until there is a sign of recovery. Lay the casualty in the 'recovery position'.

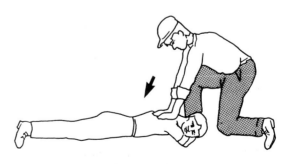

position, electrically operated plant and machinery should have a means of local isolation provided, readily identifiable and suitably placed, to give protection to persons carrying out mechanical or electrical maintenance.

In certain situations where it may be necessary to disconnect the supply rapidly to prevent or remove a hazard, a means of emergency switching should be provided. This usually takes the form of emergency stop buttons, which should be easily accessible and suitably identified. When these are operated they should not further increase the risk of hazard. All means of protection, switch fuses and isolators should be clearly marked. They should be capable of being locked in the off position so that they cannot be operated without the knowledge of persons working on that particular circuit. If the means of isolation is by fuses then these should be placed in the pocket or locked away, fuses should **never** be removed or replaced without first turning off the supply.

Before turning off a circuit, a check should be made to see that this will not cause problems for people in other parts of the building; computers, for example, can lose all their data if the supply is turned off unexpectedly. If the type of work you are doing exposes current-carrying cables or equipment to touch, and other people could come into contact with them, then you should erect barriers to prevent this and notices should be placed in position indicating the dangers.

Testing that circuits are no longer live

It is the electrician's or instructed person's responsibility to ensure that when an isolator is operated then the circuit has been cut off from the supply. The test for this should be carried out by the use of an approved tester; that is to say one specifically designed for the job and not a lamp holder and flex, or even a neon tester which can be unreliable. Before putting the tester into use, it should be established that it is functioning correctly by testing it on a known source of supply. A typical approved tester is shown in Fig. 3.3. There are a number of different types of these available, and although they are not cheap they could save someone's life.

Reduced voltage tools and equipment

It is desirable (Health and Safety at Work etc. Act Rule 35) that where at all practical, tools and equipment used on site should be of the reduced voltage type. These are normally of the 110 V a.c. (55 V to earth) type and are readily available. Care should be taken of the equipment and the following safety rules should be observed:

- Do not use power tools on lighting circuits;
- Never carry power tools by their cables;
- Check power tools for damage regularly;
- Check cables for damage, especially where they enter equipment;
- Keep plugs and sockets clean and in good order;
- See that tools are fitted with the correct plug.

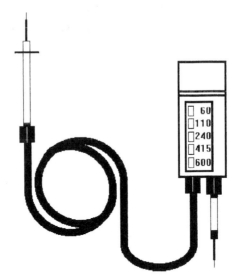

Fig. 3.3 Approved voltage tester

The drawings in Information Sheet No. 3D show a typical 110 V transformer and plug, and a circuit diagram which shows the earthing arrangement.

The need for effective earthing

Many of the electric shocks that have been recorded in recent years could have been avoided if the electrical system concerned had been effectively earthed. We shall look at the subject of earthing in greater detail later; however, it will be worthwhile at this stage to examine earthing in the context of protection against electric shock.

If a fault was to occur between the phase conductor and the metalwork or exposed conductive parts of a piece of electrically operated equipment and a person was to touch it they would receive an electric shock. The intensity of the shock would depend on a number of factors, the health of the person receiving the shock, how well they were insulated from earth, i.e., if they were wearing wellingtons or standing on a wooden floor. Contact with water lowers the resistance of the body, so wet or damp situations such as bathrooms or showers are potentially dangerous. We have seen from the drawing on Information Sheet No. 3A that when the current passes through the person receiving a shock, it travels through the earth via the fault path back to the supply authorities' transformer. If we were to provide an alternative path for the fault current to pass through, then any fault current which flows in the metalwork would be conducted safely to earth. It is for this reason that the earthing of electrical installations by a system of protective conductors is carried out, and why it is so important that this is maintained in an effective condition.

Information Sheet No. 3D Reduced voltage supplies.

1. The BS 4343 plugs
   and sockets are used
   on construction sites
   for the distribution of
   electrical supplies. The
   110 V plug and socket
   which is coloured yellow
   will be most familiar
   to the electrician.

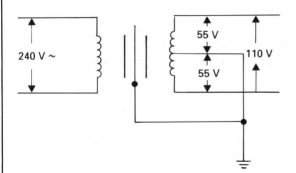

2. The 240 a.c.
   supply is transformed
   down to 110 V by
   the type of
   transformer shown.

3. This transformer
   ensures that only
   a maximum of
   55 V to earth can
   be obtained if you
   come into contact
   with live parts.

3.3 Fire prevention

Conditions required for combustion

Before any fire can exist, three conditions have to be satisfied (see Fig. 3.4). If any one of the three is missing, then the fire cannot be sustained and will go out. The three conditions are as follows:

Fuel   Any combustible material.

Heat   Hot enough to ignite the material.

Oxygen   Found in the atmosphere.

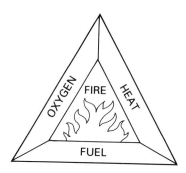

Fig. 3.4 The three conditions for fire

When a fire does break out, it can cause loss of life, the destruction of buildings and the loss or damage of valuable articles. It makes sense therefore to try and prevent fires before they start. Most fire prevention schemes are based on making sure that all three of the above conditions do not exist at the same time. Listed below are some of the **Dos** *and* **Don'ts** associated with fire prevention:

Do	Don't
Do pay attention to no smoking signs;	Don't discard cigarette ends or matches carelessly;
Do store inflammable substances in approved stores or compounds;	Don't place inflammable materials near heat;
Do pay attention to warning signs about naked flames;	Don't put hot ashes or caustic materials in waste bins;
Do turn off heating appliances after use;	Don't leave bonfires unattended;
Do regularly check equipment using bottled gas for leaks;	Don't overload electrical circuits;

Do	Don't
Do ensure all electrical connections are secure;	Don't use equipment with damaged or worn flexes;
Do remember some substances give off vapours which can ignite;	Don't be careless when using blowlamps;
Do allow sufficient ventilation around equipment you install;	Don't fit equipment without backplates on inflammable surfaces.

Methods and equipment for extinguishing fires

The methods and equipment used to fight fires differ in accordance with the type of substance that is ignited. In Table 3.1 there are some of the more common substances that you would be likely to meet and the type of equipment you would use to put out the fire.

Table 3.1

Extinguisher	Colour	Type of fire	Remarks
Carbon dioxide	Black	Electrical, oil	Causes less damage to electrical equipment
Dry powder	Blue	Electrical, fires in petrol and diesel engines	
Water type	Red	Paper, wood, rags and over-heated gas cylinders	Do not use on electrical or petrol fires
Foam or solvents etc.	Cream	Burning liquids: petrol, oil, paint	Will not spread the fire
Halogenated hydrocarbon, BCF	Green	Electrical, petrol and paint, etc.	Do not use in confined spaces

Dangers associated with fire fighting

It is essential that the correct extinguisher is used when fighting fires, as the use of the wrong type can make matters worse. Water should not be used on petrol or oil fires, for example, as this tends to wash the liquid into new areas, so spreading the fire. Do not use water on electrical fires as water will conduct electricity and you could receive a shock.

Smaller fires can be put out by the use of sand or a fire-proof blanket, while in larger installations fires would be put out by the use of automatic systems, such as sprinklers, or the filling of a room with an inert gas. Whichever system is used, you should be aware of it and be able to act accordingly.

Remember that some materials, and in particular plastic materials, give off toxic fumes which can be harmful. Beware of smoke, and if the fire is too big leave the building quickly and calmly sounding the alarm as you go. If you have time, see that doors and windows are closed as this robs the fire of oxygen; however, leave them if this will endanger yourself. Most companies have fire drill procedures, and you should make yourself aware of these, and also the positions of alarm points, extinguishers and fire exits.

3.4. Fundamental requirements for safety

From the very early days of electricity there has been an essential requirement for electrical installations to be installed safely, as well as being suitable for the purpose for which they were designed. The IEE Wiring Regulations give the basic requirements to achieve this in Chapter 13 of the Regulations and briefly they are as follows:

- Good workmanship and proper materials shall be used throughout the installation;
- The installation and equipment shall be installed in such a way as to be accessible for testing, inspection and maintenance as far as is practical;
- All equipment shall be suitable for the maximum power demand of the equipment when it is functioning in its intended manner;
- Electrical conductors shall be of sufficient size and current-carrying capacity for the purpose for which they are intended;
- Electrical conductors shall be insulated, protected and installed, so as to prevent danger as far as is practical;
- Joints and connections shall be properly constructed, regarding conductance, insulation, mechanical strength and protection;
- Where necessary, circuits will have suitably rated automatic protective devices for protection against overcurrent;
- Whenever the prospective earth fault current is insufficient to operate the above, a residual current device shall be fitted;
- Electrical equipment shall be earthed in such a manner that earth leakage currents will be discharged without danger;
- If metal parts of other services can be touched simultaneously with the above, then they should be earthed;
- No protective device shall be installed in an earthed neutral conductor, with the exception of a linked circuit breaker;
- Single pole switches shall be inserted in phase conductors, only with the exception of linked switches;
- Circuits supplying electrical equipment shall have effective means of isolation as necessary, to prevent or remove danger;
- Safe means of access shall be afforded for persons to operate, or give attention to, installed equipment;
- Equipment exposed to adverse weather or other corrosive conditions shall be designed to prevent any danger from this;

- No additions to installations shall be made without ascertaining there is sufficient spare capacity for it;
- No additions to installations shall be made without ensuring that the earthing arrangements are adequate;
- Testing shall be carried out on completion of the installation, to the requirements of the IEE Wiring Regulations.

The above requirements are fundamental to ensuring that installations are carried out safely and are suitable for the purpose that they were installed for, and we shall be returning to them more fully in other sections of the book.

Test 3

Choose which of the four answers is the correct one.

(1) The voltage to earth of a 110 V transformer of the type used on construction sites is:

(a) 110 V;
(b) 50 V;
(c) 220 V;
(d) 55 V.

(2) The most suitable fire extinguisher for electrical fires is:

(a) Foam;
(b) Dry powder;
(c) Water type;
(d) Carbon dioxide.

(3) A person touches a live conductor. You must first:

(a) Isolate the person from the supply;
(b) Go and get help;
(c) Get a doctor;
(d) Give the person cardiac massage.

(4) The three conditions for combustion are:

(a) Fuel, heat and a naked flame;
(b) Fuel, heat and oxygen;
(c) Heat, combustible material and a naked flame;
(d) Combustible material, fuel and oxygen.

(5) The correct colour-coding for a carbon dioxide fire extinguisher is:

(a) Blue;
(b) Green;
(c) Black;
(d) Red.

Chapter 4
The Measurement, Setting Out and Fixing of Equipment

4.1 Measurement and marking out

Bench work

When prefabricating cable trunking or tray, or even the metal support brackets for these items, much of the preparation can be carried out on some form of bench, whether it be a purpose-made one or one you have rigged up temporarily. The work for the most part will consist of the measurement and scribing of lines and angles on the material prior to cutting, and this will have to be done with great care if the finished job is to be accurately assembled.

Essential tools for bench work

Below are some of the tools required to carry out this sort of work. The list shows tools which are essential for this type of work and should be in the possession of most electricians.

Scriber   A scriber is a thin steel bar ground to a fine point, used like a pencil, and is capable of marking most metals used by the electrician.

Fig. 4.1 Try square and scriber in use

Try square   When cutting trunking or tray into sections, it is essential if the work is to look right that the edges are cut square; that is to say the cross cut is at 90° with the sides of the trunking or tray. The try square enables us to carry out this task. When the square is placed on the work, make certain that it fits tight up to it and is not held off by any dirt or burrs. If this is not done, the square will not be at 90° and the line we draw with the scriber will not be true. The use of both scriber and try square is shown in Fig. 4.1.

Tape measure   For accuracy an engineer would use a steel rule; however, for most purposes, the steel tape measure is sufficiently accurate for the electrician's use. Consisting of a flexible steel tape which is usually marked in both imperial and metric units of measurement, it slides neatly into its case, which is no more than 50 mm × 50 mm in size and fits easily into the pocket. Most of these tapes have a lug at the end, which can be hooked over the ends of work to be measured; the tape is drawn tight and the measurement carefully marked off (see Fig. 4.2).

Centre punch   This is used mainly for marking the spot where the electrician wishes to drill the work, although it can be used to define a line with a row of punch marks where a piece of work is to be cut, particularly if the work is to be welded and a scribed line might be lost. Many pieces of work have been spoiled by the failure of the electrician to use this tool, as high speed drill bits tend to *wander* if the metal is not *centre popped* beforehand.

Desirable tools

Below is a selection of more sophisticated tools which, if you are required to carry out a lot of this type of work, would be found most useful.

Inside callipers   If the accurate measurement of the inside of a pipe or enclosure is required, then the inside calliper must be used. The legs of the calliper are placed inside the work to be measured and extended outwards until a sliding fit is achieved.

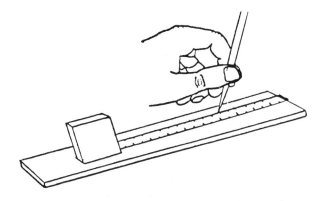

Fig. 4.2 Use of tape measure and scriber

Outside callipers   Used for accurate measurement of the outside of work, the legs are closed around the work until a sliding fit is achieved. To achieve accurate measurement with both of these callipers, they are taken to a steel rule and the measurement made.

Odd legs callipers   Consisting of one leg like an outside calliper and the other like a scriber, they are used in the opposite way to the above callipers. The measurement is taken off the steel rule first and then the calliper is placed on to the work. The calliper leg is slid along the true edge of the work and the scriber allowed to scibe a mark. Details of the above can be found on Information Sheet No. 4A.

Steel rule   For accurate measurement the steel rule is a must. It is not affected by distortion in the way a steel tape can be, and the marks showing the units of measurement never rub off. It is used with all the tools described above, and provided it is kept lightly oiled, will last a lifetime.

Bevel   A bevel or adjustable square can be most useful when marking angles on work. The engineer's version of this tool is marked off in degrees and is most accurate in use.

4.2   Tools and equipment for setting out

The positioning or *setting out* of items of equipment in their correct places on site calls for a certain amount of skill, but the task can be made a lot easier if the correct tools are employed. Below is a list of tools required to carry out this type of work:

- Spirit level;
- Straight edge;
- Plumb/chalk line;
- Try square (large);
- Water level.

Information Sheet No. 4B shows how some of these tools should be employed.
    Building is not an exact science and often you will find that rooms are not always square; that is to say adjacent walls are not always at right angles to each other. Failure to check this can mean that measurements can be out by several centimetres. If you are not in the possession of a large try square, then you can easily make a square for yourself. A set square can be cut from a single piece of plywood, making quite sure that the lengths for the sides are in multiples of three, four and five (the three, four, five triangle). This device has been used since ancient times by builders and is well worth the trouble taken to make it. Suitable lengths for the sides would be 30, 40 and 50 cm respectively. Fig. 4.3 shows the square in use.

Information Sheet No. 4A Callipers.

1. Inside callipers used for taking internal dimensions.

2. Odd legs callipers used for marking up work as shown.

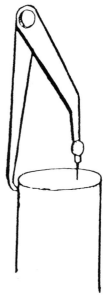

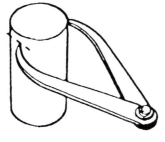

3. Outside callipers used for taking external dimensions.

Information Sheet No. 4B Setting out.

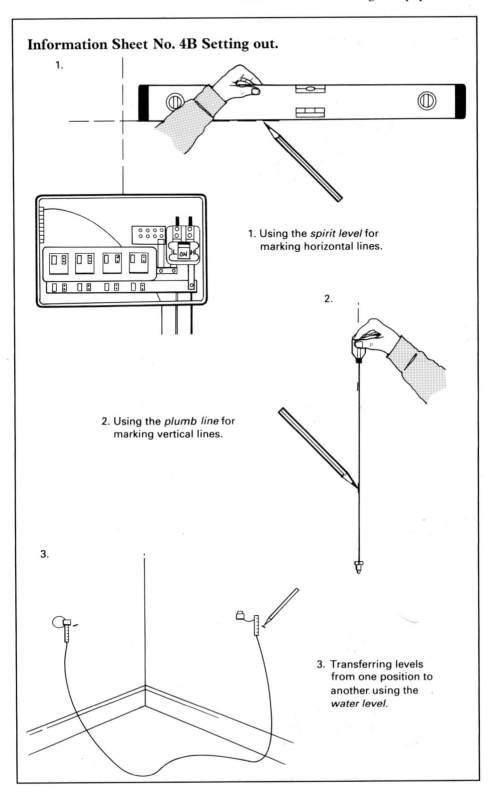

1.

1. Using the *spirit level* for marking horizontal lines.

2.

2. Using the *plumb line* for marking vertical lines.

3.

3. Transferring levels from one position to another using the *water level*.

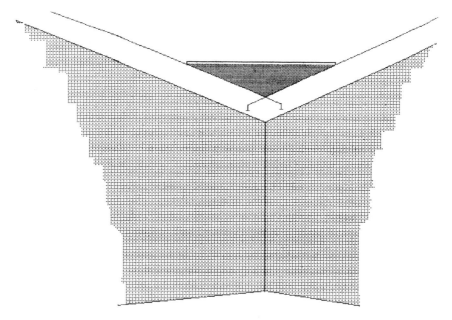

Fig. 4.3 Square in use

4.3 Fixing methods

Plastic plug and food screw

Position the article to be fixed using the marking out methods mentioned above. If the article is small, *offer it up* to the wall in the chosen position and mark the position of the fixing holes with a pencil or bradawl.

Remove the article and where the pencil or bradawl have left their mark, make a small cross with the mark at the centre. If the article is large, it will be necessary to measure the fixing centres with your rule and transfer these to the position in which you wish to fix the article. Alternatively, a template can be made from stiff card; the fixing holes are marked on this and it is a simple job to offer this up to the wall instead of the heavy piece of equipment.

Select a suitable sized screw for the size of the article you are going to fix, the bigger the article the larger the size of screw. Most electrical switch, socket and conduit boxes can be fixed with a No. 8 plug and screw. The length of the screw will to some extent depend on the weight of the article to be fixed, also if the masonry is poor you will need to drill a deeper hole to obtain a sound fixing. Use a masonry drill the same size as your choice of screw. These sometimes *wander* a little when you start to drill, but you will be able to re-centre the bit if you have placed a cross on the mark as suggested earlier. It is advisable when fixing electrical switch boxes to always use round-headed screws if the boxes are not counter sunk, otherwise there is a great risk of the conductors being cut on the sharp-edged head of the counter sunk screw. Fig. 4.4 shows the above operations being carried out.

Rawlbolt fixing

Heavy equipment requires something a little more substantial to fix it then the modest plastic plug and screw. The rawlbolt is designed to provide a quick and easy method of achieving heavy duty fixings. There are a number of variations of the fixing, but the two most commonly used are the ones offering either the stud fixing or the bolt fixing. The position of the fixings is marked as described above, and the hole is drilled to suit the size of the rawlbolt in use. The rawlbolt shell, complete with bolt, is placed very carefully into the hole and a check is made to ensure that the hole is the correct depth. At this stage, the bolt is screwed carefully out leaving the shell behind; the work is offered up and the bolt screwed back in. The rawlbolt works on the principle that as the bolt is screwed home the shell expands and grips the sides of the hole, making a secure fixing. Information Sheet No. 4C shows how the bolt fixing type is utilised.

Spring toggle

With today's modern construction methods, electricians find themselves having to fix equipment to walls and ceilings constructed of plasterboard. Normal fixing methods using plastic plugs and screws are of no use. However, a useful device that has been around for many years overcomes this problem; this is the spring toggle. A hole is made in the plaster board a little larger than the fixing bolt that is to be used (the toggle comes complete with bolt). The wings of the toggle are folded back against the pressure of a tiny spring and inserted into the hole. When the toggle goes through the hole into the cavity behind, the spring forces the wings outwards. The screw is now tightened up until the fixing is secure. Information Sheet No. 4C shows how this is done.

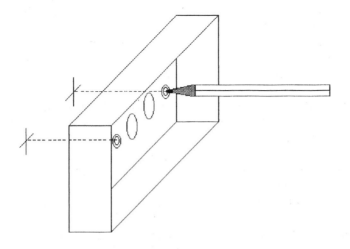

Fig. 4.4 Marking fixing holes

Information Sheet No. 4C Fixings.

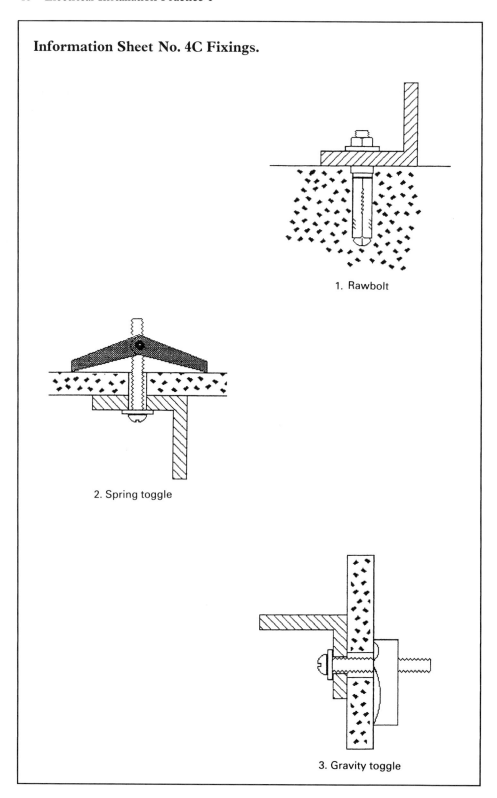

1. Rawbolt

2. Spring toggle

3. Gravity toggle

Gravity toggle

A different type of fixing for cavity walls is the gravity toggle. The hole is made in the same way as the example above and the toggle is passed through and into the cavity behind. The toggle which consists of a hollow plated steel bar mounted off-centre, drops down under the force of gravity, as shown in Information Sheet No. 4C. The bolt is tightened up and the toggle grips against the inside of the wall.

Test 4

Choose which of the four answers is the correct one.

(1) One of the tools listed below is not essential to marking out:

(a) Scriber;
(b) Try square;
(c) Tape measure;
(d) Micrometer.

(2) Which of the following is used for taking internal diameters:

(a) Odd leg callipers;
(b) Outside callipers;
(c) Inside callipers;
(d) Engineers' compasses.

(3) To ensure a conduit drop is vertical, use is made of a:

(a) Plumb line;
(b) Chalk line;
(c) Water level;
(d) Try square.

(4) A spring toggle is suitable for fixing into:

(a) High density concrete;
(b) Plasterboard ceiling;
(c) Brickwork;
(d) Ceramic tiles.

(5) Which of the following is suitable for fixing heavy equipment:

(a) Gravity toggle;
(b) Screw and plastic plug;
(c) Spring toggle;
(d) Rawlbolt.

Chapter 5
The Distribution of Electricity
to the Consumer

5.1 Local distribution of electricity

The sub-station

The distribution of electricity at local level is at 11 kV and consists for the main part of small sub-stations often connected together in the form of a ring for additional security of supply. Here the 11 kV three wire supply is stepped down by the use of a *delta-star* transformer to a 415 V four wire supply, with the *star point* connected to earth, as shown in Fig. 5.1.

Distribution of 415/240 V supplies

In the towns and suburban districts, the four wire 415/240 V supplies will be distributed in the form of an underground cable. This three phase and neutral supply, as it is referred to, will be taken directly into the premises of larger users such as small factories, blocks of offices and flats, and large commercial premises. For the smaller consumer only, one of the phases, together with the neutral, will be made available; this is known as a single phase supply (see Fig. 5.1).

The balancing of loads

In an attempt to share the load as evenly as possible over each of the three phases, the regional electricity companies will connect single phase supplies to each of the phases in turn (see Fig. 5.1). This helps to prevent any one phase being overloaded, thus keeping the size of the cable used to a minimum, reducing the current in the neutral and so keeping distribution costs down. When a three phase and neutral supply is taken into a building, this practice should be continued and the electrician should see to it that the various loads are balanced over the three phases as near as is practical.

5.2 Systems of main intake and earthing connection

The service intake position

The supply company's cable is terminated in the form of a service cut-out, as soon as practical after entering the consumer's premises. This equipment must contain

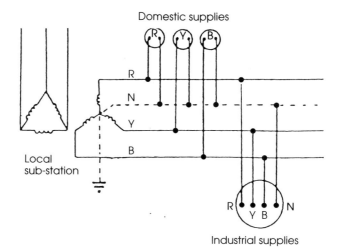

Fig. 5.1 Local distribution

provision for automatic disconnection of the supply under fault conditions as follows:

(1) Overcurrent (overload or short circuit);
(2) Dangerous earth fault currents.

Protection against these is usually provided in the form of high breaking capacity fuses (HBC), the rating of which will depend on the type of premises being supplied and the size of the service cable. Regulation 4 of the Electricity Supply Regulations 1988 requires one conductor of an a.c. distribution network to be solidly earthed and therefore this earthed conductor, or neutral as it is known, terminates in a simple terminal block or a removable solid link (see Fig. 5.2).

Methods of system earthing

Electrical supplies entering premises vary slightly with the type of earthing provision adopted. The IEE Wiring Regulations outline five different systems; however, one of these requires special authorisation before use and another cannot be used at all for public supplies. The three most popular systems are shown in Information Sheet No. 5A and it will be noted that the systems are classified with two-, three- or four-letter designation; the meaning of which is as follows.

The first letter denotes the supply earthing arrangement:

T – Earth (terre); one or more points connected to earth;
I – All parts isolated from earth or one point connected to earth through a high impedance.

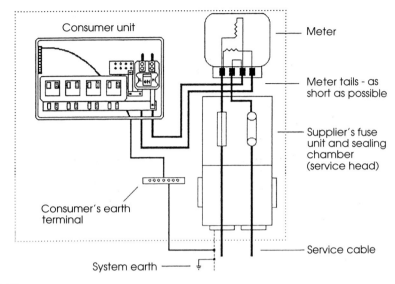

Fig. 5.2 The mains position

The second letter denotes the installation earthing arrangement:

T – All exposed conductive parts connected directly to earth;
N – All exposed conductive parts connected directly to the earthed conductor (which for a.c. is usually the neutral).

The third and fourth letters denote the arrangement of neutral and protective conductors:

S – Neutral and protective conductors separate;
C – Neutral and protective conductors combined.

Careful study of these letters and the systems shown in Information Sheet No. 5A will soon make the differences in the systems clear. Prior to the above classification system being introduced, other ways of classifying these conductors were employed and as these classifications are still used by electricity companies and other authorities it would be useful to take a look at these:

PME – This refers to protective multiple earthing (now known as TN–C–S), for which permission for its use is required from the Secretary of State for Energy in England and Wales and the Secretary of State for Scotland north of the border;
PEN – This denotes the combined protective and neutral conductors in the cables supplying the PME system;
CNE – Sometimes the above cable is referred to as combined neutral and earth;
PNB – This stands for protective neutral bonding, used on overhead systems employing individual transformers, where the supplier will, by agreement, connect an earth electrode at the consumer's premises, as well as at their transformer.

Information Sheet No. 5A Types of earthing systems.

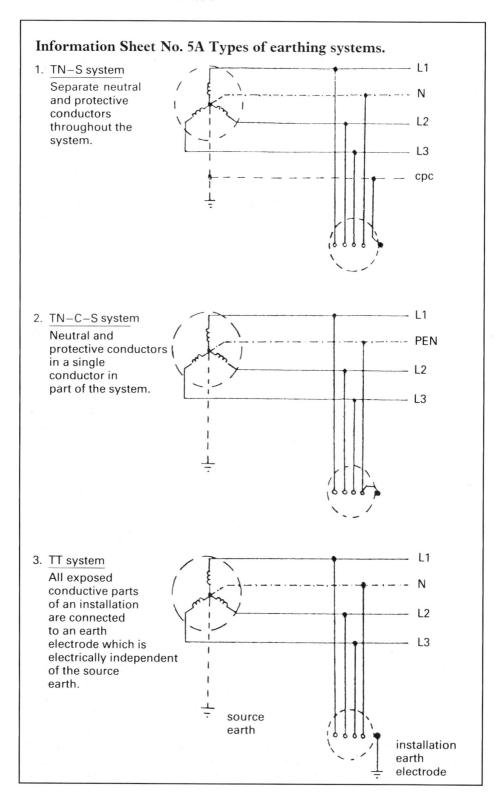

1. TN–S system

   Separate neutral and protective conductors throughout the system.

   L1
   N
   L2
   L3
   cpc

2. TN–C–S system

   Neutral and protective conductors in a single conductor in part of the system.

   L1
   PEN
   L2
   L3

3. TT system

   All exposed conductive parts of an installation are connected to an earth electrode which is electrically independent of the source earth.

   L1
   N
   L2
   L3

   source earth

   installation earth electrode

Whichever system is provided, it is the responsibility of the consumer, or the electrical contractor working for him, to satisfy themselves that the characteristics of the earth fault current path, including any part of that path provided by the supply undertaking, are suitable for the effective operation of the earth fault protection chosen for the installation.

The metering of supplies

The electric meter records the amount of energy used in a particular installation; it is connected as shown in Fig. 5.2. It will be seen that the meter has a current coil and a voltage coil; the interaction of their magnetic fields causes a disc to spin. This, in turn, is connected to gears which operate the display, which can be either the dial or digital type; the speed of the disc increases with an increase in load.

Tariffs

The unit of measure is the kilowatt hour and is the result, for example, of a load of 1 kW being used for one hour or 2kW being used for a half hour. At present, a number of different tariffs are offered by each supply company involving various methods of metering. Some of the different arrangements are shown in Information Sheet No. 5B. To calculate the amount of money owing in a domestic installation, the supply company simply deducts the previous reading on your meter from the present one. This gives them the number of units used and this is multiplied by the cost per unit, to arrive at a total unit cost. To this is added a standing or quarterly charge for provision of the service, the total being the amount to be paid. Special tariffs can be offered to some consumers where consumption can be taken in off-peak periods. An example of this is the *Economy 7* scheme for domestic users. In the meter, a time-switch operates alternative sets of dials, or switches between separate meters, thus recording energy used at both off-peak and normal times. To encourage people to use off-peak electricity, specially reduced tariffs are offered with savings of approximately 50% on the normal tariff.

5.3 Control and protection for the consumer

Switching

Every electrical installation must be provided with a main switch and be protected against short circuits, overloads and dangerous earth fault currents. Very often the functions of switching and protection are incorporated in the same device, i.e., a switch fuse. There are a number of ways that switchgear can be arranged before being connected to the supply company's meters and some of these are shown in Information Sheet No. 5B. For the most part, domestic installations will achieve compliance with the Regulations by use of a consumer control unit made to BS 1454, which combines a double pole switch rated at 60/80 A with the required

Information Sheet No. 5B Metering and switchgear layouts.

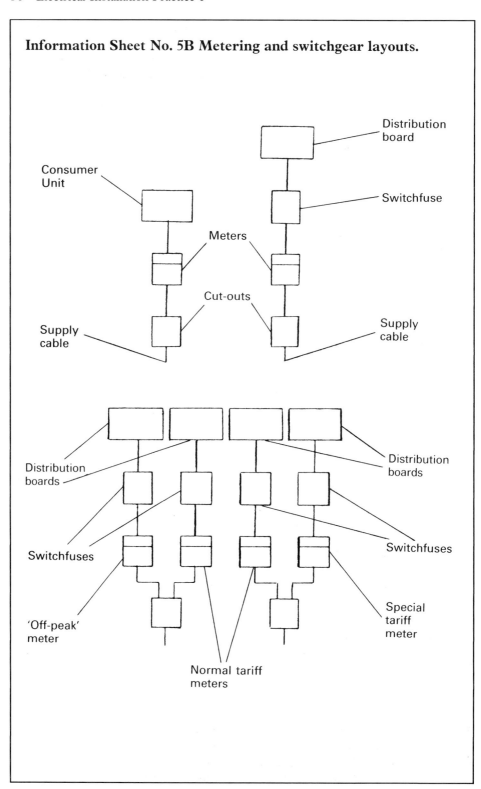

number of final circuit fuseways. Care must be taken when installing the insulated version of the consumer control unit as they often have no backs and must be mounted on non-flammable material; they can, however, be fitted with non-flammable backplates. For further information, see Information Sheet No. 5C.

Protective devices

Protection against short circuits, overloads and dangerous earth fault currents can be obtained by use of the following:

(1) Semi-enclosed rewirable fuses to BS 3036;
(2) Cartridge fuses to BS 1361 and 1362;
(3) High breaking capacity (HBC) fuses to BS 88;
(4) Miniature circuit breakers (MCB) to BS 3871.

Each of the above types of protection are of different construction and this is shown in Information Sheet No. 5D. They also have different advantages and disadvantages, and it would be useful to look at some of these characteristics here.

Semi-enclosed rewirable fuse   The advantages of these fuses are:

- They have a low initial cost;
- The fuse element is cheap to replace;
- They have no moving parts so there is less maintenance necessary;
- They are easy to check to see if they are intact.

The disadvantages of these fuses are:

- Incorrect fusewire can be used;
- They deteriorate with age;
- They cannot be replaced quickly;
- They lack discrimination;
- They are not advisable for short circuit protection.

Cartridge fuse   The advantages of these fuses are:

- They have a more accurate current rating;
- There are no moving mechanical parts;
- They are not prone to deterioration;
- They are of a small physical size.

The disadvantages of these fuses are:

- They are more expensive to replace than rewirable fuses;
- They can be shorted out with silver foil (BS 1361 type);
- The BS 1361 type can be replaced with higher rated fuse;
- They are unsuitable for very high fault current conditions.

Information Sheet No. 5C Fixing consumer units.

1. Moulded plastic units should have only enough of the KOs removed to enable the cables to enter freely. Metal units should have the metal KOs removed and bushes fitted.

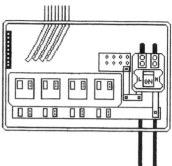

2. Fix the consumer unit to the wall with an approved fixing; strip the cables to within 2 cm of the cable entry and connect the CPCs having fitted green and yellow oversleeve.

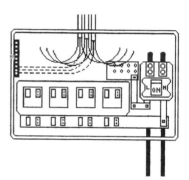

3. Next connect the neutral conductors. It is important that these are in exactly the same order as the phase conductors are going to be.

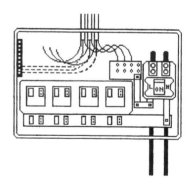

4. Finally connect the phase conductors making sure that they are in the same order as the neutral conductors. At all times leave plenty of slack cable.

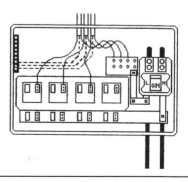

Information Sheet No. 5D Protective devices.

1. High breaking capacity
   fuse (HBC).

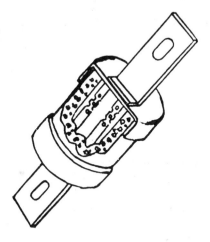

2. Miniature circuit
   breaker (MCB).

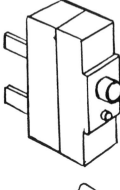

3. Rewirable fuse
   (semi-enclosed).

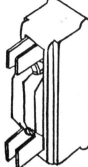

HBC fuse   The advantages of these fuses are:

- There are no moving parts to go wrong or maintain;
- There is no deterioration of the fuse element;
- They are reliable;
- The fuse discriminates between transient and sustained overloads.

The disadvantages of these fuses are:

- They have a high cost;
- It is not always easy to see if the fuse has *blown*.

Miniature circuit breaker   The advantages of these fuses are:

- They are set to a predetermined rating at the factory;
- It is easy to check if the breaker has tripped or not;
- The supply to the circuit is easily reinstated;
- Multi-pole units are available;
- They discriminate between sustained and transient overloads.

The disadvantages of these fuses are:

- They have a high cost;
- They have mechanical moving parts;
- Tripping heavy overloads causes distortion due to heat;
- Ambient temperature affects their characteristics;
- Regular tests are required to ensure their satisfactory operation.

Further work on the above is detailed in *Electrical Installation Practice Book 2* of this series.

Test 5

Choose which of the four answers is the correct one.

(1) The standard voltage between phases of a three phase supply is:

(a) 230 V;
(b) 240 V;
(c) 415 V;
(d) 500 V.

(2) The double pole switch in a consumer unit provides:

(a) Overload protection;
(b) Short circuit protection;
(c) Isolation;
(d) Correct polarity.

(3) The earthed neutral in the supply authorities' cut-out is:

(a) In the form of a solid link;
(b) Fitted with a single pole switch;
(c) Fitted with a protective device;
(d) Left disconnected.

(4) The letters PME stand for:

(a) Protective mains earthing;
(b) Protective multiple earthing;
(c) Protective mains electrode;
(d) Protective meter earth.

(5) In a domestic premises, charges are usually based upon:

(a) Kilovolt–ampere hours;
(b) Joules per hour;
(c) Coulombs per hour;
(d) Kilowatt hours.

Chapter 6
The Installation of Wiring Systems

6.1 The installation of PVC/PVC wiring systems

Reasons for choice

Polyvinyl chloride is tough, cheap, easy to work with and install, so it is not surprising, therefore, to find that the PVC insulated/PVC sheathed cable is the most popular type of cable in current use. This form of insulation has its limitations in conditions of excessive heat or cold, and can be subject to mechanical damage unless given additional mechanical protection in certain situations. However, provided this and the other factors mentioned later are taken into consideration, the PVC/PVC wiring system is probably the most versatile of all the wiring systems.

Stripping

This type of cable comes in various forms and the more common types are shown on Information Sheet No. 6A. The stripping of PVC/PVC cables is generally done with a knife. The knife should be sharp and held at a very acute angle to the cable, so that the insulation is pared rather than cut. To cut at a near angle would be very bad practice because, although it may give a neater finish, it would probably result in the nicking of the insulation or conductor. A nicked conductor becomes so weak that, after having been bent a few times, it will almost certainly break. Apart from this tendency to break, the effective cross-sectional area of the conductor will be reduced, causing increased resistance which may result in excessive heat. Techniques used in the stripping of these cables in this way are shown in Information Sheet No. 6B.

Terminating

The entry of a cable end into an accessory is known as the *termination* of the cable. In the case of a stranded conductor, the strands should be twisted together with pliers before terminating. Care should be taken that this is not overdone, as it may result in damage to the conductors. The IEE Wiring Regulations require that a cable termination of any kind shall securely anchor all the wires of the conductor and shall not impose any appreciable mechanical stress on the terminal or socket.

A termination under mechanical stress is liable to disconnection. When current is flowing, a certain amount of heat is developed, and the consequent expansion

Information Sheet No. 6A Types of cable.

1. PVC insulated
   PVC sheathed
   two core and CPC.

2. PVC insulated
   PVC sheathed
   two core.

3. PVC insulated
   twin bell wire.

4. PVC/SWA/PVC
   armoured
   cable.

Information Sheet No. 6B Stripping PVC cables.

1. Nick the cable at the end with your knife and pull apart as shown.

2. When the required length has been stripped, cut off the surplus sheathing with the knife as shown.

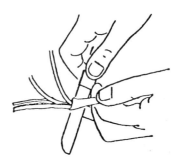

3. The insulation can be stripped from the conductors with the knife as shown.

4. An alternative method of stripping the insulation from the conductors is with a pair of purpose-made strippers as shown.

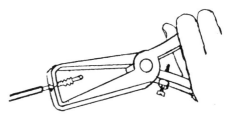

and contraction may be sufficient to allow the conductor under stress, particularly tension, to be pulled out of the terminal or socket.

If one or more strands, or wires, are left out of the terminal or socket, there would be a reduction in the effective csa of the conductor at that point. This would mean increased resistance and probably overheating; in addition to this, any strands left sticking out would be potentially dangerous. Typical types of terminal used in electrical installation work are shown in Information Sheet No. 6C.

The cable sheath of this type of wiring system is provided to give the cable mechanical protection. It is important, therefore, that this protection is maintained throughout the installation, by ensuring that the sheath enters the accessory or enclosure. If, for any reason, termination to a piece of equipment requires removal of this sheath ouside the enclosure, then attitional mechanical protection should be provided.

There are a number of different forms of this type of cable, many of which include a circuit protective conductor (CPC). This CPC is not insulated and therefore when cables of this type are terminated, the protective conductor should be sleeved with the appropriate sized yellow-and-green striped PVC over the sleeve. Further details of the identification of conductors can be found in the IEE Wiring Regulations 514–06–01 and Table 51A.

Installation

The general installation of these cables should present no difficulty, provided the above factors are taken into consideration, and the techniques used are shown in Information Sheet No. 6D. When cables are installed above ceilings or under floors, they should be run in such a way as not to be damaged by floor or ceiling boards or their fixings (IEE Regulations 522–06–05). If the cable cannot be run through joists at a distance of 50 mm or more from the top or bottom of the joist, then they should be mechanically protected in such a way as to prevent nails, screws and the like penetrating them. Where these cables are installed under cement or plaster, they should be contained in a suitably bushed piece of conduit, or alternatively covered with a piece of metal or plastic channel. Where cables enter metal enclosures, the entry should be protected by a suitable grommet to prevent abrasion of the cable sheath and the possible fault to earth. If cables pass through walls, these should be made good with suitable non-combustible material to prevent the spread of fire.

6.2 The installation of PVC/SWA/PVC cables

Reasons for choice

We have seen in Chapter 3 that all cables to be installed underground shall have built into them a sheath or armouring able to resist mechanical damage; the PVC insulated steel wire armoured PVC sheathed cable is just such a cable. This cable comes in a number of forms and one of the most popular types is shown in Fig. 6.1, together with the cable gland used for terminating the cable. Its use is not

Information Sheet No. 6C Types of terminals.

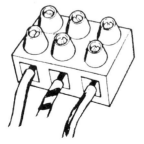

1. Typical PVC
   insulated
   connector block.

2. Stud terminal
   using a nut and
   washers.

3. Screw and washer
   connection.

4. Terminal post
   connection.

5. Claw washers used mainly
   on light conductors or flexibles.

Information Sheet No. 6D Installation of PVC cables.

1. Typical examples of clips used to fasten PVC cables. They come in all sizes to suit the different cables and cross-sections.

 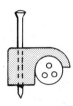

2. The cable can be straightened by running the thumb over it before clipping. The palm of the hand can be used for the bigger cables.

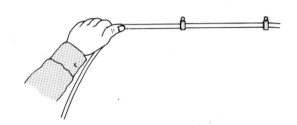

3. A cross-pane hammer will be found to be the most usful type to employ.

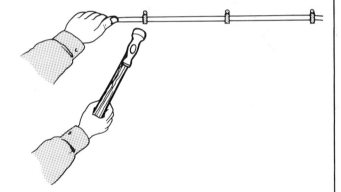

4. The clips should be placed equal distances apart and these, and the radius of the curve, can be found in the IEE Regulations.

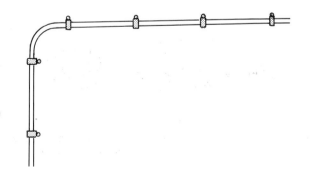

limited to underground installations, however, as the added protection afforded by the sheath makes it ideally suitable for those industrial or commercial situations where mechanical protection is of paramount importance.

Terminating

The cable consists of PVC insulated conductors with an overall covering of PVC. Between this covering and the outer PVC sheath is embedded the galvanised steel wire armouring. The armouring is used as a protective conductor and special glands are employed to ensure good continuity between this and the metalwork of the equipment to which we are connecting. These glands vary a little from one manufacturer to another and their design also depends on the environment in which they are to be used. Details of how to terminate the cable, using a gland designed for indoor use, are given in Information Sheet No. 6E.

Start by measuring the length of armouring required to fit over the cable clamp; note the measurement. Then establish how long the conductors need to be in order to connect your equipment; make a note of this. Taking the cable, measure from the end of the cable marked A in drawing No. 2 to position B; this represents the length of the conductor required. Follow this by marking the length required for the cable clamp from B to C, as shown on the drawing. At this stage, some people will strip off the PVC outer sheath; however, this is best left on as it will hold the steel wire armouring in place for you.

Next, taking a junior hacksaw, cut through the PVC outer sheath and partly through the armouring at point B; the PVC outer sheath can now be cut away as shown in drawing No. 3. Taking each strand of the armouring in turn, snap them off at the point where they are partly cut through. Then, either using the hacksaw or a knife, cut neatly round the PVC outer sheath at point C and remove the remaining piece of outer sheath; this will leave the cable as shown in drawing No. 4. The gland can now be fitted on to the cable. First slide the backnut and

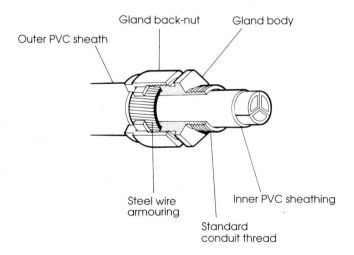

Fig. 6.1 Section of PVC/SWA/PVC cable and gland

Information Sheet No. 6E Terminating PVC/SWA/PVC cables.

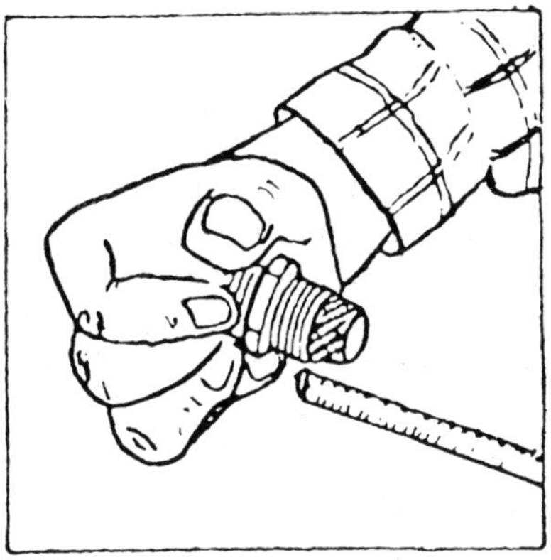

1

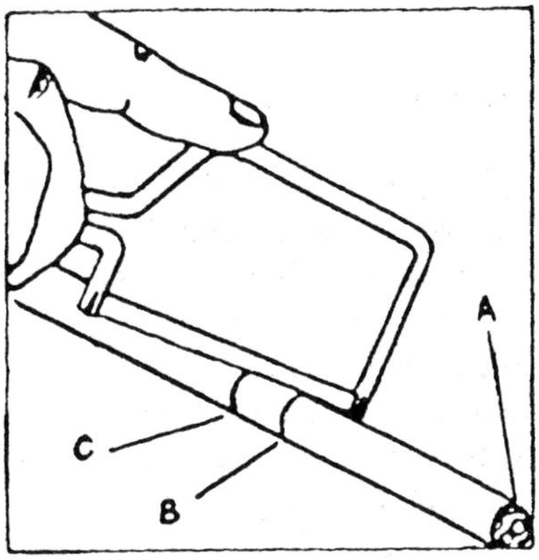

2

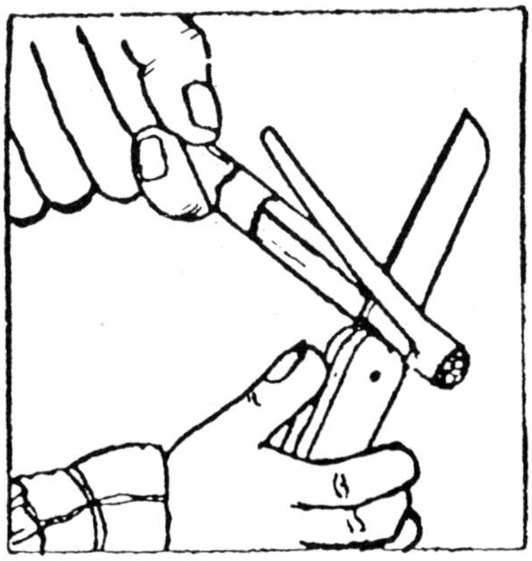

3

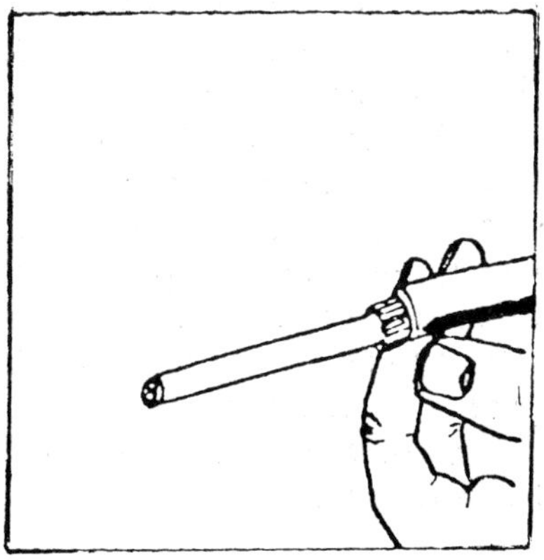

4

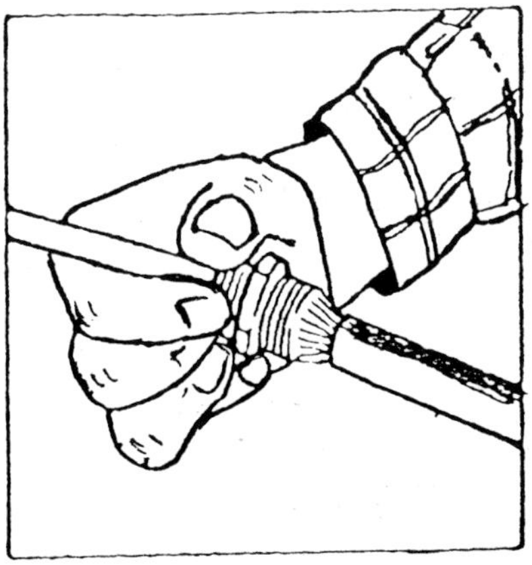

5

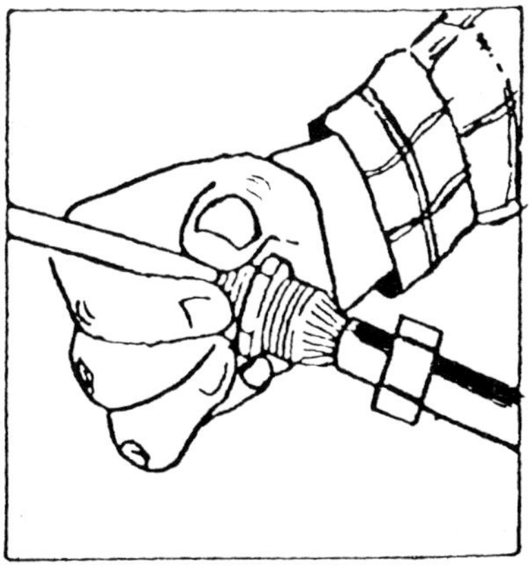

6

compression rings, if any, on to the cable. Then, taking the gland body, slide this on to the cable, making sure that it fits under all the strands of armouring as shown in drawing No. 5.

Finally, slide up the backnut and screw it on to the gland body, thus clamping the armouring tightly. The inner PVC sheath can be stripped off like any other PVC cable and the gland is ready for connection to your equipment. It is important that good contact is made by cleaning any paint work off the area of contact before tightening up the lock nut securing the gland. Bonding rings, or earthing tags as they are sometimes called, can be used to provide better contact with surrounding metalwork.

Installation

Installation is relatively easy for the smaller size cables, but it will become necessary to employ an installation team to handle the bigger sizes or multicore cables. For the most part, *one-hold* cable cleats constructed of solid PVC will be used as shown in Fig. 6.2, or in the case of cables installed on cable tray, cable ties will be utilised (see Fig. 6.3). For the bigger cables, cleats made of diecast aluminium are used. These are often designed to be slotted into steel channels so that, once a piece of channel has been fixed, multiple runs of cable can be accommodated. The minimum radius that these cables should be bent to is eight times the outside diamter, and the spacing of the cleats should be as recommended in Table 4A of the *IEE On-Site Guide*. Where these cables are to be installed directly into the ground, they should be marked with cable covers or suitable PVC marking tape to indicate their presence. They should be buried at a sufficient depth to avoid their being damaged by any disturbance of the ground reasonably likely to occur during normal use of the premises.

6.3 The installation of mineral insulated metal sheathed cables

Reasons for choice

When the ambient temperature in which cables are to be installed is high, it is difficult for the heat generated by the flow of current through the cables to be dispersed. This can result in the melting of the insulation, if adequate care is not

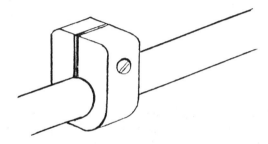

Fig. 6.2 Cable cleat

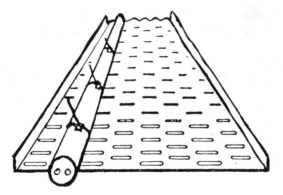

Fig. 6.3 Cable ties

taken to reduce the current-carrying capacity of the cable. PVC-insulated cables in particular suffer in this respect; however, magnesium oxide is able to withstand high degrees of temperature when used as an insulation for cables. Mineral insulated metal sheathed (MIMS) cables use magnesium oxide as their insulation, and this, together with the fact that they have a metal sheath, means that they are ideally suitable for installations where the ambient temperatures are high. These attributes are also made use of in installations where there is a high risk of fire, or in security systems, such as fire alarms and emergency lighting, where it would be desirable for the system to work for as long as possible under fire conditions.

Types of cable

The cable can be obtained with copper or aluminium sheaths; however, by far the most commonly used type is the one with copper conductors and copper sheaths (MICC). For outdoor use, or if the cable has to be run under the ground, the cable can be obtained with an over-sheathing of PVC. This PVC over-sheathing is also impervious to many oil and chemical products, so this makes it ideally suitable for installations such as petrol filling stations. There are two grades of cable: light duty rated up to 600 V is used for domestic and light duty work; and heavy duty rated up to 1000 V, which is used for industrial and other heavy duty applications. Both these cables can carry higher currents for the same cable size than other types of cable. The heavy duty cable carries a little higher current (see IEE Wiring Regulations, Tables 4J1A to 4J2B) because of its ability to disperse heat more quickly, due to its thicker outer sheath.

Termination

Magnesium oxide powder is highly hygroscopic; that is to say it absorbs moisture readily. The cable is therefore terminated using a brass pot into which is pressed a plastic compound, in order to seal the end and prevent the ingress of moisture. Should dampness enter the cable for any reason, it will only be necessary to strip back a short length of the cable to restore it to normal. If a situation arises where dampness enters the cable and it is not convenient to strip the cable back, then the pot should be removed and a blow lamp played on the end of the cable. This is most effective if the flame is started some 150 mm from the end of the cable and gradually worked to the end, thus driving out the moisture. The method used to strip and terminate the cable is shown in Information Sheet No. 6F. A number of proprietary stripping tools are now available on the market and one of the most popular types is shown in Fig. 6.4. A gland is used in conjunction with the pot when terminating the cable into equipment. It comes in three separate parts: (1) the gland body which has a standard electrical thread, usually 20 mm for the smaller cables and 25 mm for the larger; (2) an olive or compression ring, which tightens on to the cable thus affording good continuity, and (3) the backing nut which tightens down on to the olive when the gland has been positioned correctly.

Installation

The cable is very malleable and can be bent and shaped without damage; however, like all metals, if it is *worked* too much, it will eventually fracture. It is capable of withstanding hard blows and remaining operational, although mechanical protection should be provided where there is danger of damage, particularly from sharp objects. We learned earlier how *mechanical stress* can cause the breakdown of cables and that one of the causes is vibration. This can be avoided in MIMS installations if the cable is taken round in a small loop before connection to equipment, especially motors or plant likely to vibrate (see Fig. 6.5). This has the added advantage of allowing the motor or equipment to be moved a little for final adjustment, or tensioning of the belt drive.

The small overall diameter of the cable means that it can be installed under plaster without having to chase the walls, thus saving on installation costs, although PVC served cable should be used to avoid interaction between the bare copper and the finishing plaster. If the work is to be carried out on the surface, it is so neat and inconspicuous that it lends itself to the rewiring of buildings, such as churches and other historic edifices. As mentioned in Chapter 3, care should be taken to avoid situations where electrolytic action can take place and, in particular, where moisture is about.

The cable is usually fixed by the use of *one-hole* clips which can be either bare copper or PVC covered, to suit the cable being used. Other types of fixings, such as saddles, spacer-bar saddles and perforated strip, can be obtained and the cable can be fixed to cable tray by the use of cable ties in much the same way as PVC/SWA/PVC cables. The minimum internal radius that the cable should be bent to is six times the outside diameter of the cable, and the spacing of the clips should be in accordance with Table 4A of the *IEE On-Site Guide*. Identification of the conductors is particularly important in this type of wiring system, as the insulation sleeving provided is all black in colour. Polarity of the cables should be ascertained when testing is carried out and the cables marked accordingly.

Information Sheet No. 6F MIMS terminations.

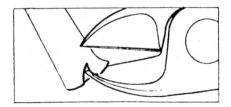

1. Start stripping
   the MIMS cable
   with the side
   cutters.

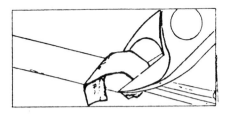

2. Curl off about
   2 cm of
   cable sheath.

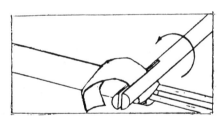

3. Insert the
   section of sheath
   into the slot of
   the stripping rod.

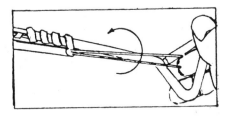

4. 'Turn off' enough
   sheath to allow
   the conductors to
   be correct length.

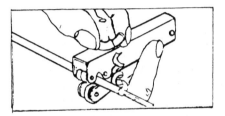

5. Place a ringing
   tool on to the
   cable and turn
   this to form a
   groove.

Information Sheet No. 6G MIMS terminations continued.

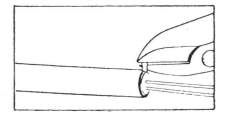

6. Snap off the surplus sheath and clean up the end of the cable.

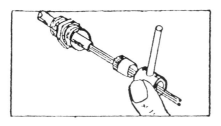

7. The cable is now ready to receive a pot using the potting tool.

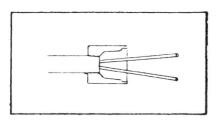

8. The end of the cable should reach the 'shoulders' of the pot as shown.

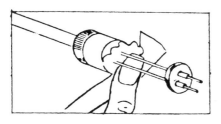

9. Fill the pot from one side using plastic compound.

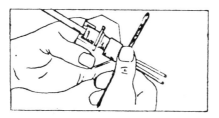

10. Finally slide on the stub cap and crimp it in position as shown.

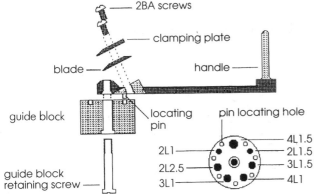

Unscrew the guide block retaining screw, recessed in the back of the guide block, in order to clear the block from the locating pin in the handle assembly. The block can then be rotated to fit the appropriate cable sizes and the locating pin engaged. *FINGER PRESSURE ONLY TO BE USED.* Do not force the guide block by tightening the retaining screw. Ensure no dirt is blocking the locating hole. Then tighten the retaining screw and the tool is ready. Do not adjust the blade.

Fig. 6.4 Stripping tool for MIMS cable

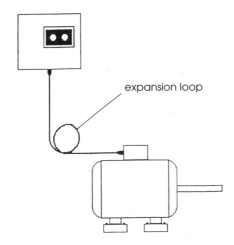

Fig. 6.5 Expansion loop in MIMS cable

6.4 The installation of conduit wiring systems

Steel conduit and the reasons for choice

Steel conduit is one of the most popular systems of wiring for commercial and industrial premises and has a number of advantages over other wiring systems. It affords excellent mechanical protection of cables, so it is entirely suitable for installations in production plants, workshops and those types of installations where the installation is subjected to a certain amount of hard use. Rewiring of this system is reasonably straightforward and circuits can be added or removed with comparative ease. Many modern buildings are constructed of in situ concrete and the conduits can be installed in the shuttering prior to the pouring of the concrete; this results in a completely concealed installation which can be wired after removal of the shuttering. Because the system is enclosed in steel throughout, it minimises risk of fire, and if certain equipment and accessories are used, it can, in fact, be used in high risk areas. When installed correctly, with all brushes and couplings correctly tightened, the steel conduit can be used as the protective conductor for the circuit, thus saving on cable.

Disadvantages of the steel conduit are that it is quite expensive compared with some other wiring systems; it requires a certain amount of skill in its installation and is therefore labour intensive. It is liable to corrosion when exposed to acid or alkaline conditions and, under certain conditions, moisture can form on the inside of the conduit, resulting in corrosion or dangerous fault conditions.

Types of steel conduit

The most popular finish for steel conduit is black enamel; however, it can be obtained in galvanised finish which makes it suitable for outdoor use, or situations subject to dampness such as dairies or bottling plants. Conduit with a coating of PVC on the outside surface can be ordered specially for use in situations subject to corrosion, although this is not seen so much since the production of plastic conduit, which we will deal with later.

The steel conduit used on the majority of installations at the present time is of the *heavy gauge* type; this comes in two variations, either the *solid drawn* or the *welded* type. The solid drawn conduit is extruded in the form of a continuous seamless tube, while the welded type is made from sheet steel rolled into a tube and welded along the seam. This welding is done so well that it is hard to tell the difference between this and the solid drawn type; however, a slight ridge can be felt on the inside of the welded type. The solid drawn conduit is more expensive than the welded type and its use is therefore limited to special situations such as gas-proof or flame-proof installations. Both types come in sizes of 16, 20 and 32 mm diameters, and in 3.75 m lengths.

Over the years, *light gauge* conduits have appeared in various forms; at the present time, their use is limited to the protection of sheathed cables from mechanical damage. They are either round or oval in shape and are formed from sheet steel, which is not welded at the seam, and they are not regarded as being suitable for use as protective conductors.

Steel conduit installations

The conduit wiring system consists of two distinct parts: (1) the conduit itself, and (2) the conductors or cables which it contains. The IEE Wiring Regulations make it quite clear that any conduit that is to be installed in situ must be fully erected before any cables are drawn into it, so it would be useful to look at this aspect of the installation first.

Installation of the conduit

First and foremost, the conduit is a metal enclosure for the protection from mechanical damage of the cables to be installed in it, therefore it should be installed in such a way as to afford continuous protection for the cables, and allow the safe and easy installation or withdrawal of such cables. Secondly, because the mass of steel used in the construction of heavy gauge conduit is of sufficiently low resistivity to comply with the recommendations contained in Regulation 543–02–04 of the IEE Wiring Regulations, it can be used as a protective conductor. It is because of this that it is essential that joints formed by couplings, bushes or accessories in a conduit wiring system are carried out in such a manner as to ensure mechanical and electrical continuity throughout its length. It is these two main points that concern us when considering the installation of steel conduits, so it may be useful to examine some of the requirements for these to ensure a safe and sound installation.

- Conduit ends should be cut squarely; use a pipe vice similar to the one shown on Information Sheet No. 6H;
- Any burrs should be removed either with a round file or a reamer, like the one shown on Information Sheet No. 6H;
- They should be threaded correctly, using stocks and dies; see Information Sheet No. 6H;
- The radius should be bent not less than 2.5 times the diameter of the conduit;
- Limited use should be made of solid elbows or tees (see Regulation 522–02–03);
- All entries into enclosures should be correctly bushed;
- Correct space factors should be applied to the number of cables installed;
- Recommendations regarding fire barriers should be taken into consideration;
- Unused conduit entries should be blanked off;
- Drainage holes should be provided to avoid collection of condensation;
- Conduits should be fixed in accordance with Table 4A of the IEE Regulations;
- All covers and box lids should be in place and securely fastened;
- All recommendations regarding corrosion should be taken into consideration;
- All bushes, couplings and accessories should be securely tightened.

Further details on the preparation and installation of steel conduits can be found on Information Sheets Nos 6I and 6J.

Drawing in of cables

When the conduit installation is complete and, in the case of a concealed system, fully dried out, the *drawing in* of the cables can begin. In a fairly long conduit run, the drawing in process is generally started at about the centre, thus reducing to a minimum the length to be pulled through.

A steel or nylon draw-in tape is first pushed through the conduit between the various draw-in boxes; a draw wire is attached to the tape and pulled through. Finally, the cable is firmly joined to the draw wire (using two loops) and drawn in (see Fig. 6.6). It is necessary for someone to feed in the cables while they are drawn in, to prevent them chafing against the edges of the boxes and to avoid them crossing each other. If there are already existing PVC cables in the conduit, care must be taken to avoid the new cables from rubbing on these, as this can cause friction burns. It is a good idea if a large number of drums of cable are involved,

Information Sheet No. 6H Conduit tools.

1.

1. Reamer. Used to clean out 'burrs' from inside of steel conduit.

2.

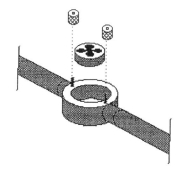

2. Stocks and dies. Used for threading the steel conduit.

3.

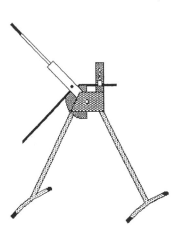

3. Bending machine. By changing the formers, these machines can be used to bend all the popular sizes of conduit.

4.

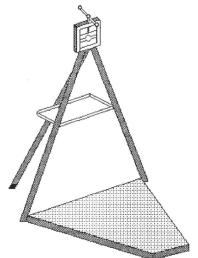

4. Stand vice. A vice which is specially designed for holding pipe or conduit while cutting or threading. The jaws are curved to help prevent the work slipping.

Information Sheet No. 6I Bending conduit.

1. Measure from the fixed point to the back of the bend as in 'd'. Place in the bending machine as shown. Using a square, line up the marked position with the edge of the former.

2. Bring the arm of the machine down checking from the side until the required angle is reached.

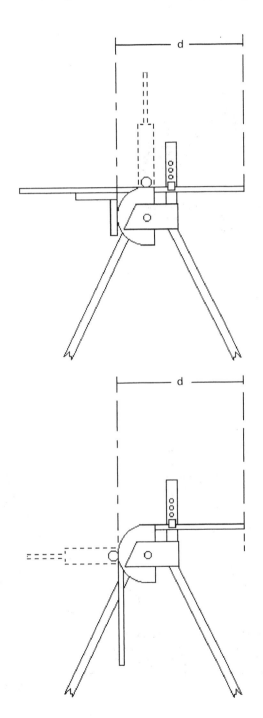

Information Sheet No. 6J Terminating steel conduit.

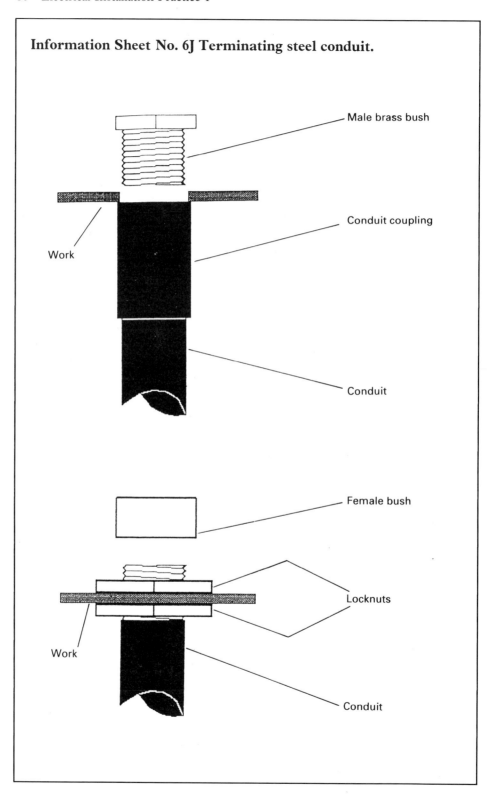

to provide some kind of stand or support, so that the cable can be pulled directly off the drums and not allowed to spiral off, so causing twists in the cable (see Fig. 6.7).

6.5 The installation of PVC conduits

Reasons for choice

Plastic conduit – usually either PVC or polythene – can be obtained in various grades and in the same sizes as steel conduits. It has many advantages over steel conduit, such as absence of condensation, elimination of abrasion and resistance to corrosion by many chemicals, it does not require painting and has excellent fire resisting properties. It is very easy to handle being light in weight, and cutting and cleaning the ends is a simple operation.

Plastic conduits do have some disadvantages when compared with steel conduits. Although they are extremely tough, they cannot stand up to the same rigorous treatment that steel conduits can. They expand very quickly when exposed to heat, so this must be allowed for when installing them in situations where this may arise. To achieve a neat appearance, they require more saddles on long runs or sagging may take place. Plastic conduits cannot, of course, be used as protective conductors, so these will take the form of additional cables and will have to be taken into account when working out space factors.

Types of plastic conduit

Super high impact   This type comes in heavy gauge and light gauge. The heavy gauge is suitable for use in severe weather conditions, while the light gauge is used for in situ concrete work.

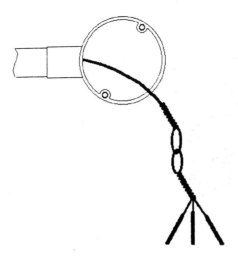

Fig. 6.6 'Drawing in' cables

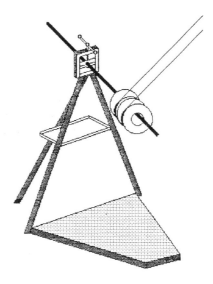

Fig. 6.7 'Running off' cables

Standard impact   Probably the most popular of the plastic conduits, it is used on installations where there are no special requirements for weather or excessive heat.

Heavy gauge high temperature   Where plastic conduits are to be installed in conditions of high ambient temperature (80–85°C), then this conduit should be used.

Flexible conduit (BS 4607)   This is available in both heavy gauge and light gauge. The heavy gauge, which is available in 25 m lengths, is suitable for either sunk or surface work. The light gauge is similar to the above, but has thinner walls.

Oval conduit and capping   Oval conduit and capping is mainly used in domestic installation, to afford protection to cables buried under plaster.

Installation of plastic conduits

Conduits made of PVC are easy to work with compared with the steel type. They can be cut by the use of a junior hack-saw held in the hand and do not require the use of a vice. Burrs are easily removed by inserting the end of your sidecutters into the tube and twisting. It is so light that it can easily be installed by one person working on their own. All the types of conduit boxes and accessories manufactured for steel conduit are available for PVC conduit, although, of course, they are also made from PVC. Plastic conduits are not threaded in the same way as steel ones, so there is no need for stocks and dies, etc. Jointing is a push fit and the end of the tube is coated with PVC solvent adhesive before inserting into the accessory. Care should be taken with this, as if it is done badly surplus adhesive may be squeezed into the tube and form a bridge, making it difficult to draw your cables in later. Make sure the adhesive is spread all round the end of the tube, or water may enter into the conduit and cause problems. A good way to ensure an even distribution of adhesive is to twist the accessory box round, so spreading it evenly.

PVC conduits of 20 mm and 25 mm sizes can be bent with the use of a steel spring, in much the same way that plumbers bend their copper pipes. The spring is inserted into the conduit with a slight twist in the direction that tightens the spring. When you have positioned it as the spot at which you wish to make the bend, it is placed across the knee or, in the case of a young person, across the thigh as their knee makes too tight a bend. Both hands should pull evenly and the bend should be taken past the angle which you wish to obtain, as PVC conduit has a tendency to spring back a little after you take the pressure off. If the tube is of greater length than the spring which you are using, tie string or a piece of cable on to it before making the bend; this will help you recover the spring afterwards, see Fig. 6.8. It may be necessary to heat the PVC conduit a little in cold weather to assist bending. Rubbing vigorously with a dry cloth is usually sufficient to produce enough heat to do the job, although hot water or the heat from a radiator can both be used to good effect. Special heaters are now on the market for use with PVC conduits and are most useful especially for the larger sizes of conduit.

We have seen that plastic conduits are subject to expansion if installed in high ambient temperature; where this is unavoidable, certain techniques can be used to minimise the problem. In long runs of conduit in these conditions, expansion couplers can be used. These should be lightly greased to keep them watertight and fitted with one coupler for every 6 m of run. More saddles are used than with steel conduit and these should be a slide fit to allow expansion to take place without buckling. The joints in high ambient temperature conditions should be made with a non-setting adhesive, which as well as allowing expansion makes a weatherproof joint. As the plastic softens with heat, care should be taken if mounting luminaires on to accessory boxes, especially the types which generate quite a bit of heat. In such cases, the temperature should not exceed 60°C and the luminaire should not weigh in excess of 3 kg.

Safety precautions should be taken with the adhesives, as they are highly inflammable and should not be left near heat. They should be stored in a place

Fig. 6.8 Bending PVC conduit

which is not accessible to young children, as they can be the subject of abuse in the form of glue sniffing.

6.6 Flexible conduits

Reasons for choice

There are many types of flexible conduits available at the present time, but they all have one common purpose and that is to provide a flexible connection between our permanent installation and some form of equipment. There are two main reasons for wanting a flexible connection: (1) to allow movement to take place of the equipment which we are connecting up, for example, to allow an electric motor to be moved on to its slide rails, in order to tighten up a vee belt drive, or (2) to prevent the transmission of vibrations to other parts of the plant.

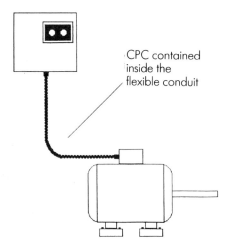

CPC contained inside the flexible conduit

Fig. 6.9 Flexible conduit to motor

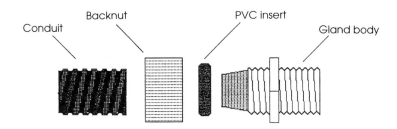

Conduit   Backnut   PVC insert   Gland body

Fig. 6.10 Connector for flexible conduit

Types of flexible conduit

Metallic flexible conduit   This consists of a light gauge galvanised steel strip spirally wound and, to some extent, interlocked, so as to form a tube. It is made in all the popular sizes and can be obtained in both watertight and non-watertight versions.

Plastic flexible conduit   This consists of a steel wire spiral covered overall by PVC. This forms a corrugated effect and makes the conduit very flexible indeed. Plastic flexible conduit too is available in all the popular sizes and is obtainable in 50 m coils.

Reinforced flexible conduit   This is a heavy duty double walled conduit with spiral wire reinforcement. It can be obtained in a number of different types to suit various applications.

Installation

Metallic flexible conduits are connected to equipment by means of special adapters, which are internally rifled at one end so as to screw on to the steel spiral; at the other end of the adapter is a standard conduit thread. Although made of steel, metallic conduits are not regarded as being suitable to act as protective conductors, so therefore an additional cable must be installed to act as a CPC in accordance with Regulation 543–02–01 (see Fig. 6.9).

Terminating plastic flexible conduits is a simple operation, provided that the correct adapter is obtained for the type of flexible conduit you are using. These often have a coloured PVC insert, which is a different colour for each type of conduit. Most of the adapters work on the principle of inserting a slightly tapered piece into the conduit's inside, then screwing on a backnut which effectively grips the flexible conduit firmly. Care should be taken when cutting the conduit to ensure a clean cut, otherwise difficulty may be experienced with inserting the tapered piece into the flexible conduit. A clean cut will also make it easier to slip the backnut on to the conduit before you start the operation. A typical flexible conduit adapter is shown in Fig. 6.10.

Test 6

Choose which of the four answers is the correct one.

(1) The letters PVC stand for:

(a) Plastic vinyl covering;
(b) Poly vinyl chloride;
(c) Poly vinyl covering;
(d) Poly vinyl chlorine.

(2) The cable sheath of PVC insulated and sheathed cable is to:

(a) Provide mechanical protection;
(b) Keep the conductors together;
(c) Give a neat appearance;
(d) Prevent corrosion.

(3) MIMS cables absorb moisture easily. This attribute is termed:

(a) Hydrochloric;
(b) Hygroscopic;
(c) Hydropathic;
(d) Hygrometric.

(4) When installing flexible conduit to equipment:

(a) The cables should be pulled in together;
(b) The gland should be tight for good continuity;
(c) A separate CPC should be installed;
(d) The end of the conduit should be square.

(5) Before pulling cables into the conduit the conduit should be:

(a) Given a coat of paint;
(b) Checked to see that it is level;
(c) Have its continuity checked;
(d) Completely erected.

Chapter 7
Installing Lighting and Small Power Circuits

7.1 The installation of lighting circuits

Final circuits

Electrical apparatus is connected by cables to the electricity supply, and to the associated protective and controlling devices (usually fuses and switches). This arrangement of cables is known as a *circuit* and circuits which connect current, using apparatus to the consumer unit or distribution board, are called *final circuits*.

Lighting final circuits

One of the earliest commercial uses for electricity was for the lighting of premises; indeed, some of the early installations had only lighting installed, as the number of electrical appliances were few.

The simplest lighting circuit is one lamp controlled by one switch and is known as a one-way circuit (see Fig. 7.1). The circuit commences at the protective device in the consumer unit, which is connected to the phase conductor of the supply. From here it goes to the switch controlling the circuit and from there to the lamp. From the lamp the cable returns to the consumer unit where it is connected to the neutral terminal of the consumer unit, so completing the circuit.

When additional lighting points are required, it would be very wasteful to connect each lighting point by its own cables to the consumer unit, therefore the original circuit is extended as shown in Fig. 7.2. This circuit now has two lamps, both of which are controlled by one switch.

If the lamps were required to be switched independently from each other, it

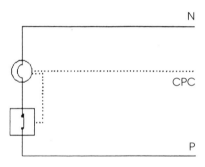

Fig. 7.1 One-way lighting circuit

would be necessary to extend the circuit as shown in Fig. 7.3. It will be noted that in all cases the circuit protective conductor (shown in dotted lines) is connected to all switch and lamp positions; this is a requirement of the IEE Wiring Regulations.

Two-way lighting circuits

For independent control from two positions, for example on a staircase, two-way switches are required. These switches have three terminals, one of which is called the *common* and is marked with a letter C; the other two are called the *strappers* and are usually marked L1 and L2 respectively. A simple two-way circuit is shown in (a) of Information Sheet No. 7A. It will be seen that the neutral conductor is taken to the lamp position. From the other side of the lamp, a conductor known as the switch wire is taken to the common of the second switch, and the two switches are linked by a pair of conductors known as the strappers. From the common of the first switch, a conductor known as the switch feed is taken to the phase.

With the switches in the positions shown in drawing (a), the current travels from the common of the first switch across the switch contacts to L2. From L2 it travels along the L2 strapper to the L2 terminal of the second switch; here it cannot go any further because the contacts of the second switch are open, so that the lamp does not light.

To make the lamp light, it would be necessary for someone to operate switch one so that the common is in contact with L1, as shown in drawing (b), or to operate switch two so that its common was in contact with L2. Either of these actions would complete the circuit and the lamp would light.

Intermediate lighting circuits

If it is desired to have control from three or more positions, *intermediate* type switches are necessary as well as the two two-way switches. Intermediate switches have four terminals and although the switch action of different makes of switch end up with the same results, the connections vary, so it is advisable to check the switch action before connecting up. The circuit is wired as shown in drawing (c)

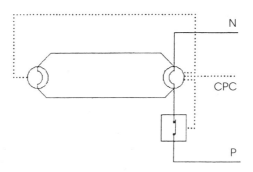

Fig. 7.2 Two lamps controlled by one one-way switch

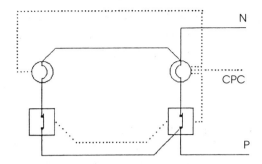

Fig. 7.3 Two lamps controlled by two switches

of Information Sheet No. 7A; it will be seen that the intermediate switches connect to the two strapping cables. This means that the circuit must always start and finish with the two-way switches. When using the commonest type of intermediate switch for three-way control, the circuit is wired as shown. The switch action in position one is shown with the solid line, and in position two with the dotted line. Operation of the two-way switches is carried out as normal and the lamp can be turned on or off from any of the three positions.

In another often used type, the L1 strappers from the two-way switches are connected into the nearest of the top terminals, but the L2 strappers are taken into the furthest of the bottom terminals (i.e. the cables are crossed over). The switch action in position one is shown and the light is off; position two is shown with a dotted line and in this position the light would be on (see drawing (d)).

Conversion of a one-way circuit into a two-way circuit

On occasions, the electrician is called upon to make alterations to existing circuits. One of the more popular requests is to make a one-way circuit into a two-way. The conversion can be carried out quite simply by running a piece of three core and CPC cable from the existing switch position to the new position. The connections are made as shown in Fig. 7.4.

7.2 Methods of wiring lighting circuits

The loop-in method of wiring

The circuit diagrams shown in Figures 7.1–7.4 are for circuits wired in single core PVC insulated cable and are suitable for wiring carried out in conduit wiring systems. Much of the wiring done today, however, is carried out in composite cables such as PVC insulated, PVC sheathed, twin and earth cable. The technique used for this type of cable is essentially different for that of the singles cables, and the first method we are going to look at is the *loop-in* method.

This is probably the most common method of wiring domestic premises in use

Information Sheet No. 7A Two-way and intermediate circuits.

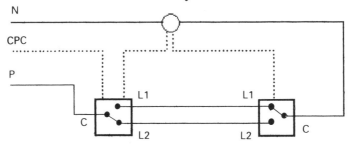

(a) Two-way circuit position one

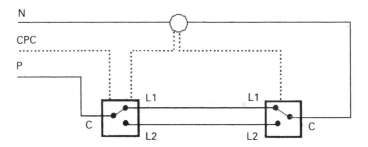

(b) Two-way circuit position two

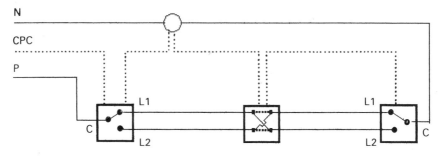

(c) Intermediate circuit type one

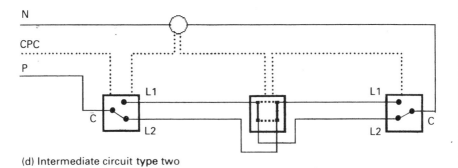

(d) Intermediate circuit type two

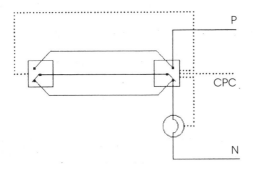

Fig. 7.4 Conversion of a one-way lighting circuit into a two-way lighting circuit

today. All the connections are made at the electrical accessories. A cable containing phase neutral and CPC conductors is run from the consumer unit to the first lighting point; a second cable is run down to the switch position. The connections are made inside the ceiling rose at the terminals provided and it should be noted that it is a requirement of the IEE Wiring Regulations that the phase terminal in the ceiling rose shall be shrouded. The reason for this is that, even with the switch in the off position, this terminal is still live until the power is switched off at the consumer unit.

If a further lighting point is required, an additional cable is run from the first lighting point to the new position. The phase, neutral and CPC conductors are connected into the corresponding connections on the first ceiling rose. At the new position, another ceiling rose is fitted and a cable taken down to the new switch position. The connections at the second position are made off in exactly the same way as before, see Fig. 7.5. This procedure is known as looping in and out of the accessories, hence the name *loop-in system*. If any of the current carrying conductors are coloured black, then they must be identified with a red sleeve or piece of red tape, both at the ceiling rose and at the switch.

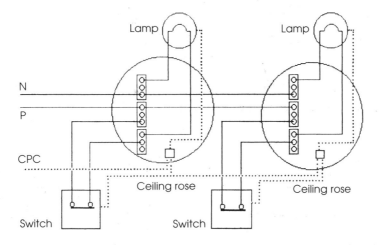

Fig. 7.5 'Loop in' method of wiring

The joint box method of wiring

There are a number of different types of joint box, but the most popular pattern consists of a circular moulded plastic box in which is fixed four or more brass pillar terminals.

The joint box is sited in a position as near to the centre of the area to be wired as possible, and fixed with wood screws to a suitable timber bearer nailed between the floor or ceiling joists. A composite cable, which contains the phase, neutral and CPC conductors, is run from the consumer unit and terminated in the joint box. Care should be taken to see that the cable sheath enters into the joint box, so no conductors are exposed on the outside. The CPC conductor is bare in composite cables, so it will be necessary to insulate this from the other cables in the joint box. This is done by fitting over it a plastic sleeving, coloured green and yellow in accordance with the IEE Wiring Regulations. To complete the circuit, further cables are run from the light position and the switch position, and the connections are made as shown in Fig. 7.6.

7.3 IEE Regulations concerning lighting circuits

We have already seen a number of the regulations applicable to the installation of lighting circuits; however, there are several other points which must be noted.

- Where conductors or flexibles enter a luminaire, as, for example, when a bulkhead fitting or batten lampholder is used, the conductors should be able to withstand any heat likely to be encountered, or sleeved with heat resistant sleeving, Regulation No. 522–02–02.
- A ceiling rose, unless specially designed for the purpose, should have only one flexible cord, Regulation No. 553–04–03.

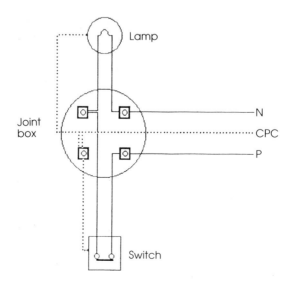

Fig. 7.6 Joint box method of wiring

- The flexible cord used to make up a pendant (the ceiling rose, flex and lampholder assembly) should be capable of withstanding any heat that is likely to be present in normal use, Regulation No. 522–02–02.
- Where a flexible cord supports or partly supports a luminaire, the maximum mass supported shall not exceed the values stated in Table 4H3A of the Regulations.
- A ceiling rose shall not be used on a voltage exceeding 250 V, Regulation No. 533–04–02.
- Parts of lampholders, installed within 2.5 m of a fixed bath or shower, shall be constructed or shrouded in insulating material. Bayonet-type (B22) lampholders shall be fitted with a protective shield to BS 5042 (Home Office skirt), or a totally enclosed luminaire installed, Regulation 601–11–01.
- Lighting switches shall be installed, so as to be normally inaccessible to persons using a fixed bath or shower. Regulation No. 601–08–01 does not apply to ceiling switches operated by an insulated cord.
- For circuits supplying equipment in a room containing a fixed bath or shower that can be touched at the same time as exposed conductive or extraneous conductive parts, the protective device shall disconnect the circuit within 0.4 of a second, Regulation No. 601–04–01.
- For circuits on TN or TT systems, where an Edison screw lampholder is being used, the outer contact shall be connected to the neutral conductor, Regulation No. 553–03–04.
- Final circuits for discharge lighting (this includes fluorescent luminaires) shall be capable of carrying the total steady current, viz the lamp's associated gear and its harmonic currents. Where this information is not available, the demand in volt-amperes can be worked out by multiplying the rated lamp watts by 1.8. This is based on the assumption that the power factor is not less than 0.85 lagging, Appendix 1 of the *IEE On-Site Guide*.
- Semi-conductors may be used for functional switching (not isolators) provided that they comply with sections 512 and 537 of the Regulations.
- When installing lighting circuits, the current is equivalent to the connected load with a minimum of 100 W per lampholder, see Table 1A of the *IEE On-Site Guide*. It should be noted, however, that diversity can be applied to lighting circuits in accordance with Table 1B of the *IEE On-Site Guide*.

7.4 The installation of 13 A socket outlets

Types of circuit

These circuits supply socket outlets to BS 1363, and permanently connected equipment, via a fused connection unit.

There are two types of circuit listed in Table 9A of the *IEE On-Site Guide* for the installation of 13 A socket outlets: (1) the radial final circuit, and (2) the ring final circuit.

(1) The radial final circuit commences at the consumer unit and loops into successive socket outlets, as shown in Fig. 7.7; the circuit ends at the last

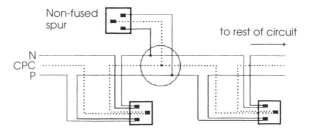

Fig. 7.7 A radial final circuit

socket. If the cartridge fuse or circuit breaker, used as an overcurrent protective device, is rated at 30 or 32 A and a minimum conductor size of 4 mm² copper, or 2.5 mm² mineral insulated cable is used, then an unlimited number of socket outlets or fused connection units can be installed in an area not exceeding 50 m². In Table 9A of the Regulations, this is classified for description purposes as the A2 circuit.

If the radial circuit is protected by any overcurrent device rated at 20 A and a minimum conductor size of 2.5 mm² copper or 1.5 mm² mineral insulated cable is used, then an unlimited number of socket outlets can be installed in an area of 20 m².

In Table 9A this circuit is classified for description purposes as the A3 circuit.

(2) Ring final circuits commence at the consumer unit and loop in and out of successive socket outlets, as shown in Fig. 7.8. The circuit does not end at the final socket as in the radial circuit, but returns back in the form of a ring to the consumer unit. Here it is connected into the same terminals from which it commenced, see Fig. 7.8. In Table 9A this is classified for description purposes as the A1 circuit.

The ring final circuit is protected by any 30 or 32 A rated overcurrent protective device, and should be wired in cables of minimum size 2.5 mm² copper, or 1.5 mm² mineral insulated cable. In an area of 100 m², or less, the number of socket outlets or fused connection units is unlimited.

Spurs

A spur is a branch cable connected to a ring or radial final circuit; see Fig. 7.8. These fall into two categories: (1) fused spurs and (2) non-fused spurs.

(1) A fused spur is connected to the circuit through a fused connection unit, the rating of the fuse in the unit not exceeding that of the cable forming the spur, and, in any event, not exceeding 13 A.
(2) Non-fused spurs feed only one single or one twin socket outlet, or one item of permanently connected equipment. Such a spur is connected to the circuit via the terminals of a socket outlet, a spur joint box, or at the origin of the circuit at the consumer unit. The size of the conductor used in non-fused spurs should not be less than that used in the ring.

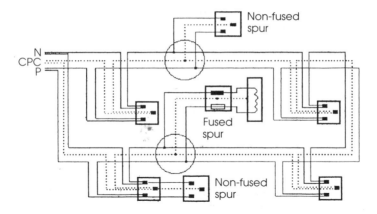

Fig. 7.8 A ring final circuit

7.5 IEE Regulations concerning 13 A socket outlets

- For the purposes of Table 9A, each socket outlet of a twin or multiple socket outlet is regarded as one socket outlet.
- Diversity between socket outlets and permanently connected equipment has already been taken into account in Table 9A and no further diversity should be applied.
- The cable sizes given are the minimum size and may have to be amended if the circuits are affected by grouping or high ambient temperature; see Appendix 4 of the Regulations.
- The total number of fused spurs is unlimited, but the number of non-fused spurs does not exceed the total number of socket outlets and items of stationary equipment connected directly to the circuit.
- The circuit protective conductor is connected in the form of a ring, as shown in Fig. 7.8, except where a metal conduit or enclosure is used as the circuit protective conductor.
- All socket outlets should be protected by a device that will operate in 0.4 seconds.
- No standard 13 A socket should be installed in a room containing a bath.
- Where equipment is to be used outdoors and therefore will be outside the equipotential bonding area, the socket outlet feeding that equipment must be protected by a residual current device with an operating current not exceeding 30 mA.
- Adjacent sockets should be on the same phase of the supply.
- All socket outlets on a TT system must be protected by an RCD to BS 4293.
- Although Table 9A is mainly intended for use in domestic type premises, it can be used for other premises, providing the maximum demand of the current using equipment does not exceed the ratings of the protective device given in the table.

Test 7

Choose which of the four answers is the correct one.

(1) A ceiling rose shall not be connected to a voltage exceeding:

(a) 250 V;
(b) 240 V;
(c) 415 V;
(d) 500 V.

(2) If a luminaire is to be controlled from three positions:

(a) A three core cable will have to be used;
(b) Three two-way switches will be required;
(c) A three gang switch will have to be utilised;
(d) An intermediate switch will have to be used.

(3) The minimum assumed demand for a lighting point is:

(a) 60 W;
(b) 100 W;
(c) 150 W;
(d) 200 W.

(4) All 13 A socket outlets should be protected by:

(a) A protective device that will operate within 0.4 seconds;
(b) A short piece of capping covering the cable connected to it;
(c) A double pole switch;
(d) A single pole switch.

(5) The number of fused spurs allowed on a ring final circuit is:

(a) 12;
(b) Unlimited;
(c) Equal to the number of sockets on the ring;
(d) Less than the number of sockets on the ring.

Chapter 8
The Inspection and Testing of Electrical Installations

8.1 Inspection and testing

Requirements of the IEE Wiring Regulations

Every electrical installation shall be inspected and tested in accordance with the Regulations, before being connected to the public supply. This is to ensure, as far as practicable, that all the requirements of the Regulations have been carried out and the installation is safe to use. The Regulations require that the tests carried out shall not in any way be a danger to persons, property or equipment, even if the circuit is faulty. It is important then that the tests are carried out in the recommended sequence, shown in Part 6 of the Regulations, and this is as follows:

(1) Continuity of protective conductors;
(2) Continuity of ring final circuits;
(3) Insulation resistance;
(4) Insulation of site built assemblies;
(5) Protection by SELV and the separation of circuits;
(6) Protection against direct contact by barrierss and enclosures;
(7) Insulation of non-conducting floors and walls;
(8) Verification of polarity;
(9) Earth fault loop impedance;
(10) Earth electrode test;
(11) Testing of RCDs.

Testing final circuits

The testing of electrical installations is an important and skilful job and is best carried out by persons suitably qualified to do so. The above tests will be covered in detail in Books 2 and 3; however, it is not beyond the scope of the beginner to carry out certain of the tests on any final circuits that they might have installed and we shall look at these here.

Visual inspection

Before any of the above tests are carried out, it is a requirement of the Regulations that the installation be visually inspected. Some of the things to look for when carrying out the inspection are as follows:

- Methods of protection against indirect contact should be in place;
- Methods of protection against direct contact should be in place;
- Connection of single pole devices for protection or switching should be in the phase conductor only;
- Selection of conductors for current-carrying capacity and voltage drop should be in accordance with the design;
- There should be a notice at the mains intake position, stating that the installation should be regularly inspected and tested;
- Switchgear should be labelled to indicate its purpose;
- There should be warning notices indicating the presence of any voltages exceeding 250 V;
- Circuit charts should be provided indicating the size of cables, protection and load;
- All cables and conductors should be correctly identified;
- Points of connection to the earth electrode and bonding should carry a notice saying 'Safety Electrical Connection – Do Not Remove';
- Equipment should be checked to see if it complies with British Standards;
- Ensure that no electronic equipment is in the circuit as this could be damaged by the test;
- Check that the protective system, including bonding, is in place as some of the tests carried out later are at mains potential and could otherwise prove dangerous;
- A check should be made to see that good workmanship and proper materials have been used in accordance with Regulation 130–01–01;
- A correct choice and setting of protective and monitoring devices should have been made;
- All equipment should be erected, installed, connected and protected to comply with the fundamental requirements for safety, as listed in Chapter 13 of the Regulations and Chapter 3 of this book.

Further details can be found in Regulation 712–01–03.

Continuity of ring final circuit conductors

A test is required to verify the continuity of the phase, neutral and CPCs of every final ring circuit.

With each leg of the ring final circuit identified, the phase conductor of one leg and the neutral of the other leg are bridged temporarily. The resistance is measured between the remaining phase and neutral conductors. A finite reading will confirm that there is no open circuit on the ring final circuit conductors under test. The remaining conductors are then temporarily bridged together.

Next the resistance between phase and neutral contacts at each socket on the ring is measured and noted. Each reading taken should be substantially the same provided there are no multiple loops existing over the length of the ring.

Where the CPC is part of the ring the test is repeated; the phase conductor of one leg and one leg of the CPC are bridged temporarily. The resistance is measured between the remaining phase and CPC conductors. A finite reading will

Information Sheet No. 8A Testing continuity of ring final circuits.

Step 1.

1. The phase conductor P1 of the first leg is bridged with the neutral conductor of the second leg N2.
The tester is then connected across conductors P2 and N1 and a finite reading confirms that there is no open circuit on the ring.

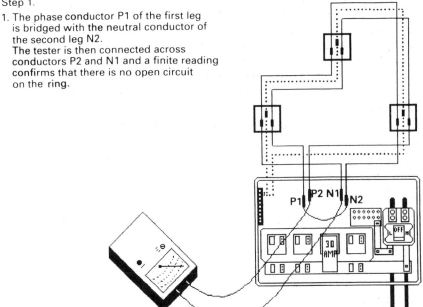

Step 2.

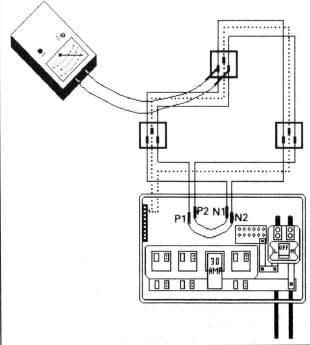

2. The next step is to bridge P1 and N2 together. The resistance between phase and neutral is now measured at each socket outlet in the circuit and a note made of the results. The results should be substantially the same provided that no multiple loops exist. Where the CPC has to be a ring, then the above tests have to be carried out by changing the CPC for one of the other conductors.

confirm that there is no open circuit on the ring final circuit conductors under test. The remaining phase and CPC conductors are then temporarily bridged together.

Next the resistance between phase and CPC contacts at each socket on the ring is measured and noted. Each reading taken should be substantially the same provided there are no multiple loops existing over the length of the ring, and the readings at the centre point of the ring are approximately equal to $(R_1 + R_2)$ for the circuit. The readings should then be recorded.

Continuity of CPCs and equipotential bonding

A test is required to verify the continuity of the CPCs including any bonding conductor.

Test method 1   Strap the phase conductor to the CPC at the distribution board so as to include all the circuit. Then test between the phase and CPC connections at each outlet in the circuit. The measurement $(R_1 + R_2)$ at the circuit's extremity should be recorded.

Test method 2   First (**Step 1**) connect one lead of the continuity tester to the consumer's earth terminal. Next (**Step 2**) use the other lead of the tester to make contact with the CPCs at various points of the circuit, such as switches, luminaires, etc.

The resistance reading will include the resistance of the test leads. The resistance of the leads should be measured and the result taken away from the first reading to obtain the final result.

To test the continuity of bonding conductors *Test Method 2* should be used.

Insulation resistance

The purpose of the insulation test is to ensure that the insulation is sound and that no faults exist between phase and neutral conductors, and between each of these conductors and earth. The test is carried out with the circuit to be tested isolated from the mains supply, using an insulation tester on which the megohm scale has been selected. The voltage used must be as shown in Table 8A.

When testing between phase and neutral, make sure that all lamps have been removed, and that all appliances are either unplugged or isolated from the circuit by switching them off. The fuses must be in place and all switches in the 'on' position (other than ones isolating appliances from the circuit). When these conditions have been satisfied, a reading is taken and this must not be less than that shown in Table 8A.

When testing between phase and earth, and neutral to earth, it is common practice to twist phase and neutral together, and test between these and earth. The instrument used is the same as above and the reading is taken in megohms. Isolate the supply as near to the mains intake position as possible; make sure the fuses are in place, and any breakers and switches in the 'on' position. When these conditions have been met, a reading is taken and this must not be less than that shown in

Table 8A Minimum values of insulation resistance

	Test volts V	Resistance MΩ
Extra low voltage circuits when supplied from a safety isolating transformer	250	0.25
Up to and including 500 V with the exception of the above	500	0.5
Above 500 V and up to 1000 V	1000	1.0

Table 8A. If a fault should be detected, it will be necessary to test between phase and earth, and neutral and earth separately, in order to ascertain which of these conductors the fault is on; see Information Sheet No. 8D.

Where equipment has been disconnected in order to carry out the tests, if practical the equipment itself must undergo an insulation test. The test result must comply with the BS standard for the equipment; if, however, there is no standard, the insulation resistance shall be not less than 0.5 megohms.

Verification of polarity

The polarity test is made to establish that:

- All fuses and single pole control devices are connected in the phase conductor only;
- The centre contact bayonet lampholder and the Edison screw-lampholder have their outer contacts connected to the earthed neutral conductor;
- Socket outlets have the phase conductor connected to the terminal marked L; the circuit protective conductor is connected to the terminal marked E, and the neutral conductor is connected to the terminal marked N.

The method for testing polarity is the same as that described for *Test Method 1* for checking the continuity of protective conductors (CPCs).

For ring circuits if the tests required by Regulation 713–03 for continuity of ring final circuit conductors has been carried out no further test is required.

For radial circuits the $(R_1 + R_2)$ measurements made in *Test Method 2* should be made at each point, and then repeated with the phase and neutral strapped together at the switchboard, and the test made between the switch line and neutral at the light point or phase-to-neutral at the equipment.

Information Sheet No. 8B Testing continuity of protective conductors.

Method 1.

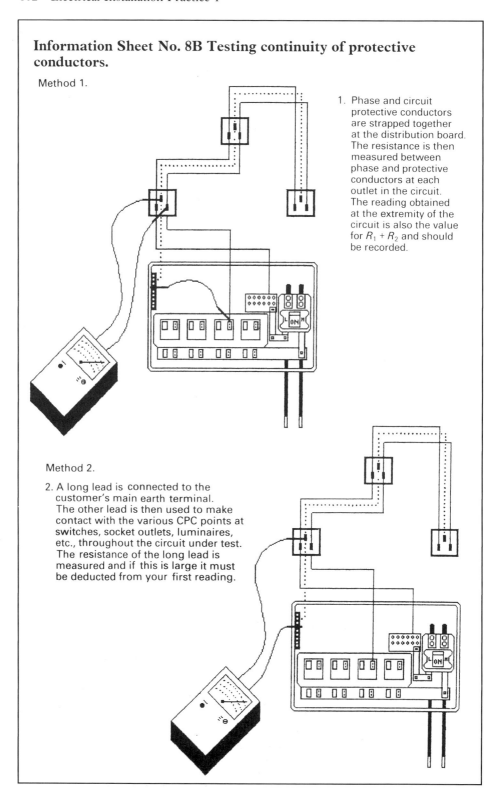

1. Phase and circuit protective conductors are strapped together at the distribution board. The resistance is then measured between phase and protective conductors at each outlet in the circuit. The reading obtained at the extremity of the circuit is also the value for $R_1 + R_2$ and should be recorded.

Method 2.

2. A long lead is connected to the customer's main earth terminal. The other lead is then used to make contact with the various CPC points at switches, socket outlets, luminaires, etc., throughout the circuit under test. The resistance of the long lead is measured and if this is large it must be deducted from your first reading.

Information Sheet No. 8C Polarity tests.

1.

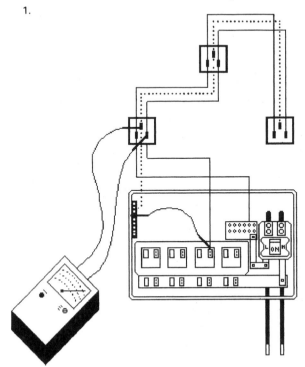

1. Phase and circuit protective conductors are strapped together at the distribution board. The resistance is then measured between phase and protective conductors at each outlet in the circuit.

2. Polarity test of an Edison screw (ES) lampholder.

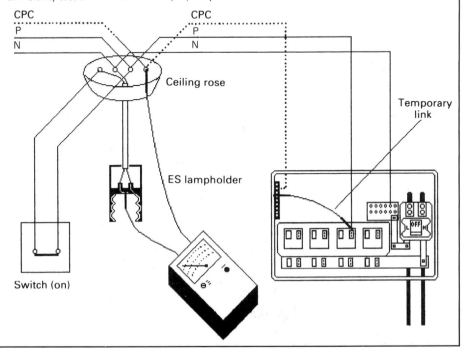

Information Sheet No. 8D Insulation testing.

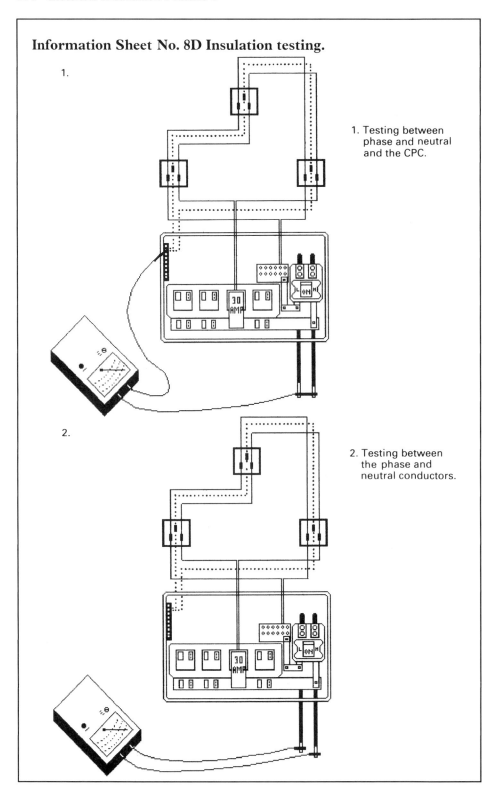

1.

1. Testing between phase and neutral and the CPC.

2.

2. Testing between the phase and neutral conductors.

Test 8

Choose which of the four answers is the correct one.

(1) The sequence of initial tests is important because:

(a) It saves time switching off the supply;
(b) Tests dependent on one another are in the correct order;
(c) It saves changing the meter over to another scale;
(d) Time is saved changing to different meters.

(2) The minimum insulation resistance on a 240 V circuit is:

(a) 0.25 MΩ;
(b) 0.1 MΩ;
(c) 0.5 MΩ;
(d) 500 MΩ.

(3) A polarity test is carried out to see that:

(a) the circuit works correctly;
(b) all current carrying conductors are coloured red;
(c) the CPC is continuous throughout its length;
(d) single pole devices are connected in the phase conductor.

(4) The continuity of a ring final circuit is checked with:

(a) an earth loop impedance tester;
(b) a high reading ohmmeter;
(c) an earth electrode resistance tester;
(d) a low reading ohmmeter.

(5) There should be a notice at the mains position stating:

(a) The installation should be regularly tested and inspected;
(b) Any voltages over 500 V;
(c) The name of the person in charge;
(d) The name of the electrical contractor.

Chapter 9
Electric Circuits

9.1 Electron flow

The structure of materials

The basic substances in our environment are known as elements and are classified according to their atomic structure. Each element has a structure which is unique to itself and all other substances are made up from combinations of two or more of these elements. Atoms are the smallest particles that can exist in nature without losing their identity as elements and can be regarded as the basic building blocks of nature.

In simple terms each atom consists of a central core which is made up of one or more protons, which carry a positive charge, and neutrons which carry no charge. This central core is referred to as the nucleus of the atom. Orbiting the nucleus are as many negatively charged electrons as there are protons in the nucleus. It is popular to picture these electrons as moving round the nucleus in much the same

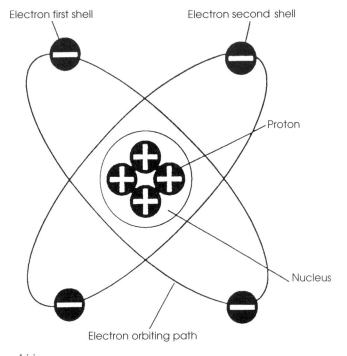

Fig. 9.1 Electrons orbiting

way that the planets encircle the Sun in our solar system. They orbit in more than one plane (see Fig. 9.1) and some electrons move closer to the nucleus than others.

The number of negatively charged electrons is normally equal to the number of positive charges in the nucleus and in this state the atom is said to be electrically balanced. However, electrons which orbit furthermost from the nucleus are not so firmly attached as those that are closer and these may more easily become detached from the nucleus. In some materials the electrons will transfer themselves from atom to atom, and an atom that is deficient in electrons is said to be positively charged, while an atom with an excess of electrons is said to be negatively charged. An atom that has lost electrons strives constantly to achieve electrical balance by attracting loose negative electrons. The movement of electrons that takes place as a result can be described in simple terms as an electron flow or electric current flow.

Materials which have loosely attached electrons are mostly metals and are good conductors of electricity. Materials whose electrons are more firmly attached and therefore more difficult to force away from the nucleus are things like mica and plastics which act as good insulators.

9.2 Production of electrical potential

Current flow

It is a basic fact of physics that energy is not lost but is simply converted into some other form of energy. Events have to follow the laws of conservation of energy, so if we need to produce a particular form of energy for our own use we will have to expend other forms of energy to produce it.

In order to make full use of the phenomenon described above it would be necessary to detach electrons in order to create the unbalanced condition necessary to cause electrons to move in a preferred direction and produce a current flow. Therefore we need to apply some other form of energy or force in order to achieve this.

There are a number of forms of energy that can be used to produce electron flow. It might be a good idea to examine in a simple way some of the ones that have particular importance in the electrical industry.

Thermal energy

When two different metals, for example copper and iron, are brought into contact with each other and heat is applied at the point of contact, then an electrical charge is produced. There are many different combinations of metals that can be used to produce this effect, though it would be foolish to utilise metals that would melt at the temperatures at which they are expected to be used.

We take advantage of the above fact in a device called a *thermocouple* where the two metals are welded together to form what is known as a junction. If the junction is connected to a sensitive meter, as shown in Fig. 9.2, and the heat applied, the circuit will act as a path for the current flow that is produced as a result of the

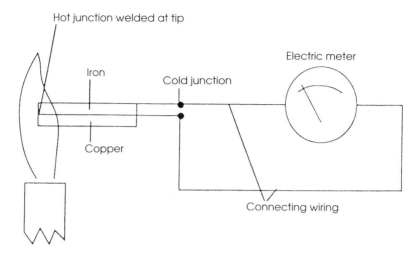

Fig. 9.2 Thermocouple

electrons attempting to restore the electrical balance to the atoms of the metal that has lost them due to the heat being applied. The resultant flow of current round the circuit will be indicated on the meter. The current is very very small and the direction of flow is from negative to positive. The heated end of the junction is known as the *hot junction* and the other end of the device is known as the *cold junction*. The greater the temperature difference between the hot and cold junctions the greater the current available.

A practical application for the use of thermocouples is in the *thermoelectric flame failure devices* used in gas- or oil-fired boilers. Here a thermocouple is placed in the flame of the boiler and connected in such a way that if the flame goes out the electrical current produced by the device no longer flows and the gas or oil supply is automatically switched off. We shall return to this topic in more detail in Chapter 11.

Magnetic energy

Most people are aware of the properties of permanent magnets and we discuss these more fully in Chapter 10. These properties are used to good effect in the electrical industry for numerous applications. For example, they are often employed to hold an appliance door shut, such as a refrigerator; and permanent magnets are used for some fractional kW motors, television scanning components, etc. How can we utilise these effects to produce electricity?

If we connect a centre scale galvanometer as shown in Fig. 9.3, and then move a section of the wire across the face of one pole of a permanent magnet, the needle of the galvanometer will move in one direction and then return to the central position. This indicates that a current was produced momentarily. Moving the wire across the face of the magnet in the other direction will cause the needle to move in the opposite direction and return to the central position, showing that a current flow occurred in the opposite direction.

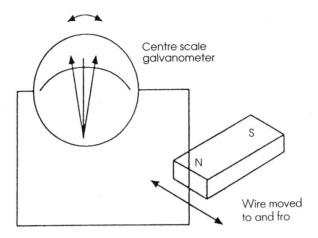

Fig. 9.3 Induction

Further experiments would show that:

- Holding the wire stationary produces no deflection of the meter and therefore no current flow;
- The direction that the current flows depends upon which direction we make the wire cross the pole;
- If we use the opposite pole of the magnet the effect of the direction of current flow is reversed;
- The faster the wire is moved across the face of the magnetic pole the greater the deflection on the meter;
- The closer we pass the wire to the magnetic pole without actually touching it the greater the deflection;
- If we increase the strength of the permanent magnet we find the deflection is increased;
- When the wire is formed into a coil the deflection increases;
- If the number of coils passing the face of the magnetic pole is increased the amount of deflection increases.

What is happening in the above experiments is that the wire which is moved across the face of the magnetic pole is cutting the lines of magnetic force produced by the magnet, causing movement of electrons in the wire and producing a current flow. The more lines of force we can cut in a given time the greater the production of electricity, and this effect is enhanced when we use a coil of wire instead of a single turn.

It was experiments like these that led Michael Faraday in 1831 to discover that an electric current could be produced by the movement of magnetic flux relative to a coil. The effect is given the name *induction* or *dynamic induction* because a current is induced into the conductor as a result of movement of the conductor or magnet relative to each other. We will return to this topic in Chapter 10.

Chemical energy

When two different metals, for example copper and iron, are placed in an electrolyte, then a flow of electrons takes place. There are many different combinations of metals that can be used to produce this effect, and indeed carbon can also be used.

We take advantage of the above fact in a device called a *simple cell* where the two metals, usually zinc and copper, are immersed in an electrolyte such as dilute sulphuric acid. The electrolyte acts chemically on the zinc and a difference in potential is produced between the two metals. If the two metals, which are known as electrodes, are connected to an external circuit, current will flow from the coper through the circuit to the zinc electrode and through the cell from zinc to copper (see Fig. 9.4).

The production of the electrical energy eventually causes the chemicals to weaken and break down and this, as well as problems with hydrogen bubbles (polarisation) forming on the copper electrode, causes the electron flow to cease.

An improvement on the simple cell is the *primary cell* which tries to overcome the problem of bubbles forming on the positive electrode. The most common of

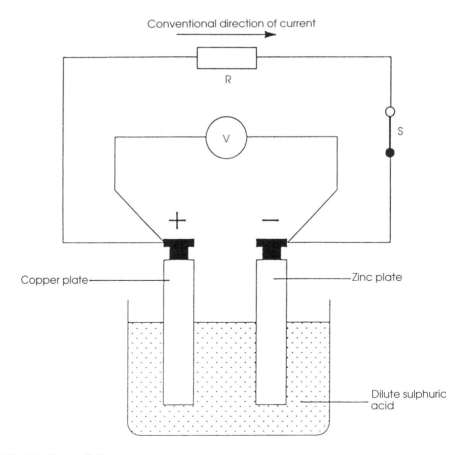

Fig. 9.4 Simple Cell

Information Sheet No. 9A Primary cells.

(a) Wet Leclanché cell

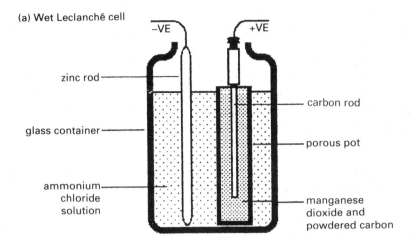

−VE

+VE

zinc rod

carbon rod

glass container

porous pot

ammonium chloride solution

manganese dioxide and powdered carbon

(b) Dry Leclanché cell

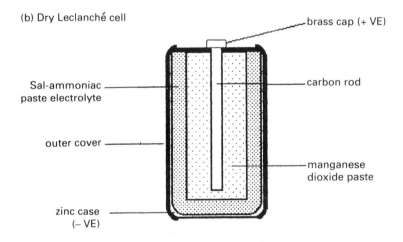

brass cap (+ VE)

Sal-ammoniac paste electrolyte

carbon rod

outer cover

manganese dioxide paste

zinc case (− VE)

(c) Mercury button cell

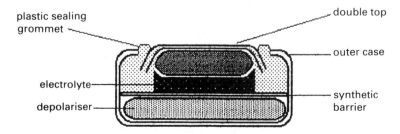

plastic sealing grommet

double top

outer case

electrolyte

depolariser

synthetic barrier

these is the *Leclanché cell*, which comes in both wet and dry types. The wet type, not used very much these days, consists of a glass container containing an electrolyte of a solution of ammonium chloride (sal-ammoniac) in water. Into this is placed the negative electrode which is a zinc rod and the positive electrode which is a carbon rod. To reduce the effects of polarisation the positive electrode is placed in a porous pot in which it is packed with a mixture of manganese dioxide and powdered carbon which acts as a depolarising agent. See drawing (a) on Information Sheet No. 9A.

The dry Leclanché cell, which is not totally dry, consists of a zinc case which also doubles as the negative electrode. The electrolyte which is in paste form surrounds the positive electrode which is a carbon rod placed in a muslin bag containing the depolariser. See drawing (b) on Information sheet No. 9A. Both the wet and dry types of Leclanché cell provide an emf of 1.5 V.

New forms of primary cell are being developed such as the *alkaline–manganese* which has a zinc negative electrode, a manganese dioxide positive electrode and an electrolyte of potassium hydroxide (alkaline). This version has up to four times the energy content of the zinc–carbon type. It has an emf of 1.5 V and is much used in personal radios, portable cassette players and children's toys.

Another recent development is the *mercury button cell*, which is much used in calculators, cameras, watches, etc. It consists of an outer sealed steel container which is the positive electrode, and a cylinder of compressed zinc powder set in the centre of the cell which is the negative electrode. The negative electrode is surrounded by an electrolyte of potassium hydroxide (alkaline) and the depolariser is manganese dioxide and carbon in pellet form. The nominal voltage for this type of cell is 1.35 V. See drawing (c) on Information Sheet No. 9A.

9.3 The electric circuit

Electromotive force

We can see from the above that by expending heat, magnetic or chemical energy, a force is provided to move the electrons in the material and cause a current flow. The force which makes this happen is called the *electromotive force* (emf) which is given the symbol E and is measured in volts (V).

It follows from this that in order for an electric current to flow in a circuit two conditions must be satisfied:

(1) There must be a source of emf to cause the movement of electrons and cause a current flow;
(2) There must be a complete path of suitable conducting material for the current to flow through.

Drawing (a) on Information Sheet No. 9B shows such an electric circuit and included is a cell for the source of supply, a switch to turn the circuit on or off and a heater which will act as a load.

Some of the terms used in electrical engineering such as current, pressure and flow can also be used for water and a hydro-analogy is often used to explain

Information Sheet No. 9B The electric circuit.

(a) Simple electric circuit

load (heater)

switch ————   ———— conductor

cell (pressure)

+   −

current flow
(conventional)

(b) Equivalent water circuit

load (radiator)

valve ———   ——— pipe

pump
(pressure)

water flow

electrical principals. Drawing (b) on Information Sheet No. 9B shows in simple terms the equivalent water circuit. This has a pump as the source for the water, a valve to turn the water on and off and a radiator to act as a load.

The coulomb

If we cause an electric current to flow in a conductor from one place to another, the amount of charge transferred is equal to the number of electrons moved. The electron is a very small quantity of electricity, however, and for practical purposes a unit called the coulomb (abbreviation C, symbol Q) is used. One coulomb is equal to 6.24×10^{18} electrons which is like saying 6.24 million million million electrons or $6\,240\,000\,000\,000\,000\,000$ electrons.

In our water analogy this would be the same as saying how many units of water we had transferred from one point to another.

The ampere

Often abbreviated to amp, this is the rate of flow of electricity and is the number of coulombs (Q) passing a particular point in a circuit in one second (t) so that:

$$\text{Amperes } (I) = \frac{Q}{t}$$

Example 9.1

If 6000 coulombs pass through a circuit in five minutes, what is the current flowing?

$$\text{Amperes } (I) = \frac{Q}{t} = \frac{6000}{5 \times 60} = 20 \text{ amps}$$

The coulomb can be regarded as an ampere second.

Applying the hydro-analogy this would be like measuring the current of water flowing by counting how many units of water had passed a particular point in one second.

Measuring current

If we require to measure the current flow in a circuit, we must arrange to count the number of coulombs passing a point in the circuit every second. In order to do this we must arrange to connect our measuring device in such a way that the current flows through the instrument. As we measure the current in amperes or amps, it follows that the instrument is called an ammeter and as the current must flow through the instrument it is connected as shown in Fig. 9.5.

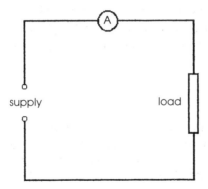

Fig. 9.5 Ammeter connection

The ampere hour

For large applications such as motor vehicle batteries, etc., the coulomb is much too small a unit and instead a unit called the ampere hour (A h) is used. One A h is the quantity of electricity passed in a circuit when a current of one amp is maintained for a period of one hour. Since there are 3600 seconds in one hour (60 × 60) it follows that there must be 3600 ampere seconds or coulombs in one ampere hour.

Example 9.2

When a current of 10 A flows in an electrical circuit, calculate the charge (a) in coulombs for 30 seconds; (b) in ampere hours for 10 hours.

(a)
$Q = I \times t$
$Q = 20 \text{ A} \times 30 \text{ s}$
$Q = 300 \text{ C}$

(b)
$Q = I \times t$
$Q = 20 \text{ A} \times 10 \text{ hours}$
$Q = 100 \text{ A h}$

Potential difference

In our water circuit there is a pump providing the driving force pushing the water round the circuit. The pump is creating a high pressure on the water, forcing it round to different parts of the system where the pressure is lower.

In our electrical circuit we have a cell providing the emf causing the current to flow round the circuit. The electrical pressure (or potential, as it can be referred to) is exerted by the source of emf and is measured in volts. Like the water circuit there is a pressure difference between different parts of the circuit and we refer to this as the potential difference or pd.

Since potential is measured in volts, pd is also measured in volts (V). It should be noted that the symbol for pd uses the sloping or italic V while the symbol V is used for the unit of voltage, e.g. 240 V, and is not italic.

Measuring potential difference

We learned earlier that pd is a name given to the difference in potential between two points in a current-carrying circuit. We saw that potential difference is measured in volts, so therefore the instrument used to measure this is called a voltmeter. Since the voltmeter measures the difference in potential between two points it follows that it should be connected across the two points, as shown in Fig. 9.6.

Resistance

The pipework of the water circuit will slow down or offer resistance to the flow of water round the system. Smaller pipes would create more resistance to the flow and the current of water would be less; larger pipes would give less resistance to the flow and this would result in a larger current of water flowing.

The conductors in the electrical circuit will behave in the same way as the pipes, smaller cables giving more resistance to the flow of electrical current and larger ones giving less resistance to the flow, with the result that a larger current would flow. The resistance (R) of an electrical circuit is measured in ohms.

Measurement of resistance

There is a requirement in The IEE Wiring Regulations to test the insulation resistance of circuits and the continuity of ring final circuits, CPCs, etc. and this

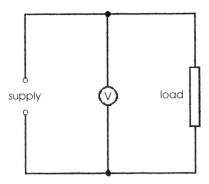

Fig. 9.6 Voltmeter connection

is discussed more fully in Chapter 8 of this book. The tests are carried out by the use of an insulation resistance tester which is an ohmmeter with two ranges. The first range tests the insulation of cables and equipment and as this will be of high resistance the scale is calibrated in megohms (MΩ), 1 MΩ being equal to 1 000 000 Ω. The second range tests the continuity of conductors and as this is of low resistance the scale is calibrated in ohms (Ω).

9.4 Ohm's law

If we wanted to increase the flow of water in our water circuit we could increase the size of the pump giving us more pressure. Provided the walls of the pipe were thick enough to withstand this increased pressure, a larger current of water would be produced quite safely.

   If we wanted to increase the flow of electrical current in our circuit we could increase the number of cells giving us more pressure or a greater potential difference between the ends of the circuit. Provided we made sure the insulation surrounding the cable was suitable for this higher voltage, then a larger current flow would be produced quite safely.

   Experiments by Dr G. S. Ohm many years ago showed that the electric current (I) flowing through a load or conductor is directly proportional to the potential difference (V) across the ends of the circuit and it follows from this that the current is inversely proportional to the resistance (R), provided the temperature does not change. The results of the experiments are known as Ohm's law and hold good for d.c. circuits and a.c. circuits containing only resistance, and can be expressed as a formula as follows:

V (volts) $=$ I (amps) $\times R$ (resistance)

   A memory aid for this formula is given in Fig. 9.7. By covering the symbol for the unit required, the formula to find the unit is given by the remaining symbols. For example, covering the symbol, R, will leave V over I, which is like saying V divided by I.

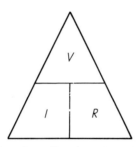

Fig. 9.7 Ohm's law memory aid

Example 9.3

In the electrical circuit shown in Information Sheet No. 9B the resistance of the heater element is 20 Ω.

(a) If the current flowing in the circuit is 5.5 A, determine the supply voltage.
(b) If the circuit is connected to a 240 V supply, what current will flow in the circuit when the switch is turned on?
(c) What value of resistance would the heater element have to be if a current of 12 A is required using the 110 V supply?

(a)
$$V \text{ (volts)} = I \text{ (amps)} \times R \text{ (resistance)}$$

therefore

$$V = 5.5 \times 20$$
$$V = 110 \text{ V}$$

(b)
$$V \text{ (volts)} = I \text{ (amps)} \times R \text{ (resistance)}$$

Transposing for *I*:

$$I = \frac{V}{R}$$
$$I = \frac{240}{20}$$
$$I = 12 \text{ amps}$$

(c)
$$V \text{ (volts)} = I \text{ (amps)} \times R \text{ (resistance)}$$

Transposing for *R*:

$$R = \frac{V}{I}$$
$$R = \frac{110}{12}$$
$$R = 9.17 \ \Omega$$

9.5 Resistivity

Conductors can be made of different materials and if we were to take a unit cube of each type of material and measure between the opposite faces of each cube we would see that each would give us a different reading. We could use the results to help us determine the resistance of a sample of that same material whatever the length or cross-sectional area (csa). Table 9.1 shows the resistivity of some of the more common materials.

The temperature would have to remain constant during such tests as an increase in temperature in normal conductive materials such as copper and aluminium results in an *increase* in resistance, although some materials used in electronic devices such as carbon and oxides of manganese do have a negative temperature coefficient.

Table 9.1 Resistivity of materials

Material	Resistivity ohm metre (Ω m)	Resistivity microhm millimetre ($\mu\Omega$ mm)
Copper	1.78×10^{-6}	17.8
Aluminium	2.82×10^{-6}	28.2
Tungsten	5.50×10^{-6}	55.0
Brass	7.20×10^{-6}	72.0
Eureka	480×10^{-9}	480.0

From the above we can say that the resistance of a conductor depends upon its length (l), its csa (a) and the resistivity of the material. This resistivity is given the Greek letter *rho* (ρ) as a symbol and the formula used to calculate the resistance of the material is:

$$R = \frac{\rho \times l}{a}$$

Example 9.4

A copper conductor has a resistivity of 17.8 $\mu\Omega$ mm and a csa of 2.5 mm². What will be the resistance of a 20 m length of this conductor?

$$R = \frac{\rho \times l}{a}$$

$$R = \frac{17.8 \times 20 \times 10^3}{10^6 \times 2.5}$$

$$R = 0.142 \ \Omega$$

Example 9.5

A copper conductor has a resistivity of 17.8 $\mu\Omega$ mm and is 1.785 mm in diameter. What will be the resistance of a 60 m length of this conductor?

We require the csa of the conductor and have been given only the diameter, so we must first find the csa.

$$\text{csa of the conductor} = \frac{\pi d^2}{4}$$

$$= \frac{3.142 \times 1.785 \times 1.785}{4}$$

$$= 2.5 \text{ mm}^2$$

Having found the csa of the conductor we can now proceed:

$$R = \frac{\rho \times l}{a}$$

$$R = \frac{17.8 \times 60 \times 10^3}{10^6 \times 2.5}$$

$$R = 0.426 \ \Omega$$

Example 9.6

A copper conductor has a resistivity of 17.8 $\mu\Omega$ mm and a csa of 5 mm². What will be the resistance of a 20 m length of this conductor?

$$R = \frac{\rho \times l}{a}$$

$$R = \frac{17.8 \times 20 \times 10^3}{10^6 \times 5}$$

$$R = 0.0712 \ \Omega$$

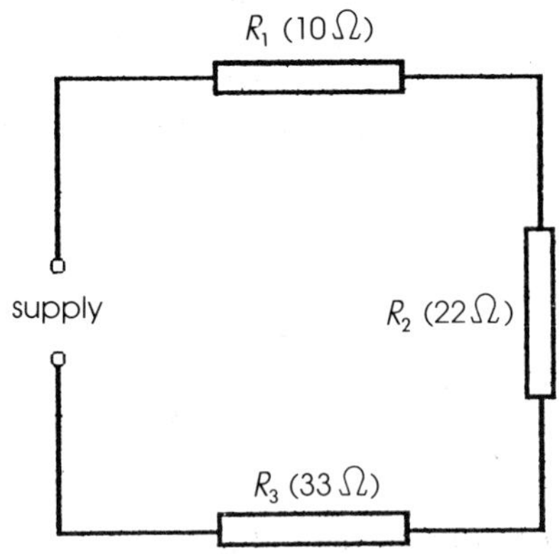

Fig. 9.8 Series connected resistors

9.6 Series circuits

Resistors in series

In Example 9.4, above, we had a 20 m length of conductor which we found had a total resistance of 0.142 Ω. In Example 9.5 we increased the conductor's length by three times and saw that the resistance also went up by three times. The fact is that when we connect any set of resistors together in a continuous line, we add their resistances together to find the total resistance. We call this connecting resistors in series.

Several resistors may be connected in series, as shown in Fig. 9.8. The total resistance of the circuit is simply the sum of all the individual resistances, so that:

$$R_t = R_1 + R_2 + R_3$$

Where R_t is the total resistance.

Example 9.7

$$R_t = 10 + 22 + 33$$
$$= 65 \ \Omega$$

Rule 1 for series circuits   When resistors are connected in series the current must flow through each resistor in order to complete the circuit and we can say that *the current flowing in a series circuit is common to all the resistors in that circuit.*

Voltage drop in series circuits

Although the current is common to each resistor in the series circuit, as we have seen above, the volt drop across each resistor will be different. The reason for this is quite simple: each resistor is of a different size and since the circuit must obey Ohm's law, i.e. $V = I \times R$, then, because each resistor is of a different value, each resistor will have a different voltage drop across it.

Let's take the three resistors used in the example above and see how this works:

Example 9.8

Three resistors of values 10, 22, and 33 Ω respectively are connected in series to a 100 V d.c. supply. Calculate:

(a) the current flowing through the circuit, and
(b) the volt drop across each resistor.

Using Ohm's law:

$$I = \frac{V}{R}$$

$$I = \frac{100}{65}$$

$$I = 1.53846 \text{ A}$$

Each individual volt drop can now be calculated applying Ohm's law:

volt drop across R_1 = $I \times R$ = 1.53846×10 = 15.38460
volt drop across R_2 = $I \times R$ = 1.53846×22 = 33.84612
volt drop across R_3 = $I \times R$ = 1.53846×33 = 50.76918

total   99.99999

It will be no surprise for us to see that the individual volt drops add up to the total applied voltage (give or take small inaccuracies).

Rule 2 for series circuits   The volt drop across each resistor will *vary*, and the sum of the individual voltage drops will add up to the total voltage.

9.7 Parallel circuits

Resistors in parallel

In Example 9.4, above, we saw that a 20 m length of 2.5 mm² copper conductor had a total resistance of 0.142 Ω.

In Example 9.6 above we doubled the csa of the conductor and found that the total resistance was 0.0712 Ω.

From this we can see that the second example is half the resistance of the first.

Remember our hydro-analogy – bigger pipe = less resistance to the flow of water? Here we have a bigger csa of cable, therefore less resistance to the flow of current. The same effect would have been achieved if we had connected two pieces of 2.5 mm² copper cable parallel with each other.

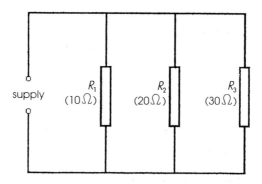

Fig. 9.9 Parallel connected resistors

When resistors are connected in parallel, as shown in Fig. 9.9, then the total resistance of the circuit can be found from:

$$\frac{1}{R_t} = \frac{1}{R_1} + \frac{1}{R_2} + \frac{1}{R_3}$$

Example 9.9

Three resistors of value 10, 20 and 30 Ω are connected in parallel. What is the total resistance of the circuit?

$$\frac{1}{R_t} = \frac{1}{10} + \frac{1}{20} + \frac{1}{30}$$

$$= \frac{6 + 3 + 2}{60}$$

$$= \frac{11}{60}$$

Inverting both sides gives:

$$R_t = \frac{60}{11}$$

$$R_t = 5.45 \ \Omega$$

It will be found that the total resistance of the circuit is always less than the smallest resistance in the circuit.

Rule 1 for parallel circuits   The supply *voltage* is common to all resistors.

The current in a parallel circuit

Although the voltage is common to each resistor in the parallel circuit, as we have seen above, the current through each resistor will be different. The reason for this is quite simple: each resistor is of a different size and since the circuit must obey Ohm's law, i.e. $V = I \times R$, then because each resistor is of a different value, each will have a different current flowing through it.

Let's take Example 9.9 above and see how this works:

Example 9.10

Three resistors of value 10, 20 and 30 Ω respectively are connected in parallel to a 20 V d.c. supply. Calculate:

(a) the total current flowing in the circuit;
(b) the current flowing through each resistor.

The total current can be found using Ohm's law:

$$I = \frac{V}{R} = \frac{20}{5.45} = 3.66 \text{ A}$$

And each individual current through each resistor can be calculated using Ohm's law:

$$\text{Current through } R_1 = \frac{V}{R} = \frac{20}{10} = 2.0 \text{ A}$$

$$\text{Current through } R_2 = \frac{V}{R} = \frac{20}{20} = 1.0 \text{ A}$$

$$\text{Current through } R_3 = \frac{V}{R} = \frac{20}{30} = 0.66 \text{ A}$$

It will be no surprise to find that if the individual current values are added they equal the total current.

Rule 2 for parallel circuits The current through each resistor will *vary*, and the sum of the individual current values will add up to the total current.

9.8 Series-parallel circuits

This type of circuit combines the series and parallel circuits, as shown in Fig. 9.10. To calculate the total resistance in a combined circuit we must first calculate the resistance of the parallel group(s) as shown in section 9.7 above. Then, having found the equivalent value for the parallel group(s), we simply add this to the series resistor(s) in the circuit to give us the total resistance for the whole of the network.

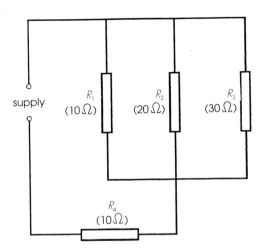

Fig. 9.10 Series-parallel connected resistors

Example 9.11

Fig. 9.9 shows a series–parallel circuit. Calculate:

(a) the total resistance of the circuit;
(b) the current flowing through the circuit.

First find the equivalent resistance of the parallel group:

$$\frac{1}{R_e} = \frac{1}{R_1} + \frac{1}{R_2} + \frac{1}{R_3}$$

$$= \frac{1}{10} + \frac{1}{20} + \frac{1}{30}$$

$$= \frac{6 + 3 + 2}{60}$$

$$= \frac{11}{60}$$

Inverting both sides gives:

$$R_e = \frac{60}{11}$$

$$R_e = 5.45 \ \Omega$$

Then we must add this equivalent value to the series resistor R_4 to find the total resistance:

$$R_t = R_e + R_4$$

$$R_t = 5.45 + 10$$

$$R_t = 15.45 \ \Omega$$

We can now use Ohm's law to calculate the current in the circuit:

$$I = \frac{V}{R}$$

$$I = \frac{100}{15.45}$$

$$I = 6.47 \ \text{A}$$

9.9 Electrical power

Electrical power is used whenever an electric current overcomes resistance. The unit of power is the watt (W) and is the power used when a pd of 1 V maintains a current of 1 A in a resistive circuit.

In all d.c. circuits and a.c. circuits containing only resistance it is possible to

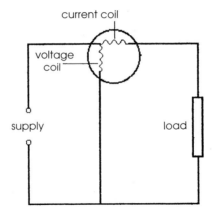

Fig. 9.11 Wattmeter connection

determine the power being used by use of an ammeter and voltmeter, as shown in Fig. 9.11 and multiplying the readings so that:

$$P \text{ (watts)} = V \text{ (volts)} \times I \text{ (amps)}$$

Power can also be measured by an instrument which is called a wattmeter. As you would expect, this consists of two coils, one a current coil which is connected in the circuit in series in the same way that an ammeter is, and one a voltage coil which is connected in the circuit in parallel in the same way that a voltmeter is (see Fig. 9.11). The interaction of the two sets of magnetic fields gives a deflection related to the power in watts.

The watt is a relatively small unit and where large amounts of power (P) are used it is more convenient to use the kilowatt (kW) where 1 kW is equal to 1000 W.

Example 9.12

An immersion heater rated at 3 kW takes a current of 13 A. What will be the pd across the element?

3 kW = 3000 W, so that:

$$P = VI \text{ therefore } V = \frac{P}{I}$$

$$V = \frac{3000}{15} = 200 \text{ V}$$

By using our knowledge of Ohm's law we can derive two more methods of calculating power:

Substituting for V we get the following:

$$P = (IR) \times I$$

$$= I^2 R$$

Substituting for I we get the following:

$$P = V \times \frac{V}{R}$$

$$P = \frac{V^2}{R}$$

Example 9.13

A circuit contains a 60 W lamp and is connected to a 240 V supply. Calculate (a) the resistance of the lamp, and (b) the current flowing in the circuit.

(a)

$$P = \frac{V^2}{R} \text{ therefore } R = \frac{V^2}{P}$$

$$R = \frac{240 \times 240}{60}$$

$$R = 960 \ \Omega$$

(b)

$$P = V \times I$$

$$I = \frac{P}{V}$$

$$I = \frac{60}{240}$$

$$I = 0.25 \text{ A}$$

9.10 Electrical energy

The basic unit of work or energy is the joule (J), but this is a very small unit which only represents a power of one watt per second. For example, a 100 W lamp requires 100 joules every second. We must therefore use a much larger unit or the figures would be unmanageable. The unit used is the kilowatt hour which represents one kilowatt for one hour. From this we can see that:

$$1 \text{ joule (J)} = 1 \text{ watt (W) for one second (s)}$$

$$1000 \text{ J} = 1 \text{ kilowatt (kW) for one second}$$

$$3\,600\,000 = 1 \text{ kW for one hour (kWh)}$$

The kilowatt hour is the unit used by the electrical supply companies to charge their customers for the supply of electrical energy. All the electric meters are rated in kWh and these are often referred to as the *Board of Trade Units* or just *Units*. Information Sheet No. 9C shows an example of how a supply company calculates the cost of electricity consumed.

Information Sheet No. 9C Specimen electricity bill showing how the cost of electricity consumed is calculated.

SOUTHERLY
ELECTRIC

Enquiries to

Manse Street
Hamton QU1 CK

Telephone No. (Mon–Fri 9.00 am–4.30 pm)
1 498 344870

TAX POINT   20 OCT 93

Mr. SMITH
10 BARN CLOSE
SOCKHAM
MIDDX 11915

REF No.   98765432

Normal Reading Date
20 OCT 93

C = Your reading	E = Estimate (please see over)		VAT No. 543211		
Previous Reading	Present Reading	Tariff	Units	Pence Per Unit	Amount £ p
22162E	23775	DOMESTIC	1613	7.490	120 81

Quarterly Charge   10   33
VAT AT 0.00% on above charges of £131.14   0   00
(TOTAL VAT THIS ACCOUNT £0.00)

TOTAL   131   14

TO PAY BY DIRECT DEBIT REFER TO LEAFLET
YOUR BUDGET SCHEME MONTHLY PAYMENT
WOULD BE   £33

Payment is now due please see over

Test 9

Choose which of the four answers is the correct one.

(1) An atom is composed of:

(a) Ions and electrons;
(b) Nucleus and one or more elements;
(c) A nucleus and one or more electrons;
(d) Protons and neutrons.

(2) A form of energy that does not produce an electron flow is:

(a) Kinetic;
(b) Magnetic;
(c) Heat;
(d) Thermal.

(3) An electrical charge can be produced by the application of heat to a:

(a) Thermostat;
(b) Thermograph;
(c) Thermometer;
(d) Thermocouple.

(4) In a d.c. circuit the current flowing through a load or conductor is directly proportional to:

(a) The resistance;
(b) The applied voltage;
(c) The ohmic value;
(d) The resistivity.

(5) The basic unit for work or energy is the:

(a) joule;
(b) watt;
(c) coulomb;
(d) emf.

Chapter 10
The Magnetic Effect of an
Electric Current

10.1 Apparent effects

We learned in the previous chapter that it is a basic fact of physics that energy is not lost but is simply converted into some other form of energy, so if we need to produce a particular form of energy for our own use we will have to expend other forms of energy to produce it. Electrical energy is no different and follows the laws of conservation of energy in that it, too, can be converted into other forms of energy.

When a current flows in a circuit, electrical energy is converted into other forms of energy and certain effects are produced. Because electricity cannot be seen, tasted or smelt we rely on these effects as evidence that electrical energy exists at all.

The three main effects of an electric current are:

(1) the magnetic effect;
(2) the heating effect;
(3) the chemical effect.

As these effects are so important we shall devote this and the next two chapters to looking at them in a little more detail.

10.2 Magnetism

The effects of magnetism have been observed for centuries. The Greeks discovered over two thousand years ago that a certain iron ore known as magnetite was able to attract small pieces of iron. The Vikings made good use of the fact that the ore lodestone, or *leading stone* as it was called, pointed in a northerly direction when suspended from a cord.

We saw in Chapter 9 that electrons move round the nucleus of an atom in much the same way that the planets encircle the Sun in our solar system. Not only do they orbit in more than one plane, but each electron spins on its axis in the same way that the earth spins on its axis. The spinning action produces a field with a magnetic polarity which is determined by the direction of spin. Non-magnetic materials like copper and aluminium have an equal number of clockwise and anti-clockwise spins so in simple terms these cancel each other out and there is no external field. On the other hand magnetic materials such as iron, nickel and cobalt have a surplus of electron spins in one direction and it is this imbalance that produces the external magnetic field.

Before we go on to explore the electromagnetic effects of an electric current it might be important to investigate the properties of these *permanent magnets* as they are called, as this is so fundamental to our understanding of the subject.

The three main materials used in the production of permanent magnets are iron, nickel and cobalt, but iron is the best and in fact magnetic materials are generally referred to as ferromagnetic materials.

These materials are often combined in alloys such as *Alnico* or *Ticonal* in order to produce particular characteristics for use in modern magnets. These are known as *hard* magnetic materials and are difficult to magnetise, but they retain their magnetism for a long while.

Magnets can be produced by stroking a piece of iron or steel (an iron alloy) with a magnet a number of times in the same direction. A much less tedious method of producing them is to place the magnetic material in a coil, through which a heavy d.c. current is passed. The magnetic force, like electricity, cannot be seen, tasted or smelt, but the magnetic field emanating from the magnetic material can be shown by placing a sheet of plain paper over the magnet and sprinkling on iron filings. If the sheet is tapped lightly the iron filings will trace out a pattern of the magnetic field as shown on Information Sheet No. 10A. These lines of magnetic force can also be plotted by the use of a compass needle, the direction of the force being taken as that to which the north end of the needle is pointing.

Further investigation of magnetism would show:

- The Earth has a magnetic field like a large permanent magnet;
- Freely suspended bar magnets set themselves up so that their axes lies in the direction of the north and south magnetic poles;
- The end of the magnet pointing north is called the north seeking pole and the other end the south seeking pole;
- *Like* kinds of magnetic pole *repel* each other;
- *Unlike* kinds of magnetic pole *attract* each other;
- The space surrounding a magnet where the magnetic effect is present is called the magnetic field of the magnet;
- It is convenient to consider that the magnetic field consists of a number of lines of magnetic flux;
- The lines of magnetic flux are concentrated at each pole and get weaker as they get further away from the magnetic pole;
- The direction of a magnetic field is shown by the direction of the north pole of a compass needle when placed in that position;
- The lines of magnetic flux never cross one another;
- The lines of magnetic flux try to form complete loops passing through the magnet;
- The part of the magnetic field outside the magnet runs from north to south;
- If a compass needle is placed in a position where it is influenced by two magnetic fields it will point in the direction of the resultant field;
- When a piece of unmagnetised magnetic material is brought into contact with a magnet it becomes magnetised itself by *induction*.

Information Sheet No. 10A Magnetic fields.

10.3 Units of magnetic flux

It is convenient to imagine that the magnetic field surrounding a magnet is permeated with lines of magnetic flux as this helps us understand the magnetic effects produced. The stronger the magnet the more magnetic flux it produces and we give this flux the Greek letter *phi* (Φ) as a symbol. The unit of magnetic flux is the *weber* (pronounced vayber) and this is abbreviated to Wb.

The concentration of magnetic flux at a particular point is called the flux density (symbol B) and is measured in webers per square metre (Wb/m²) or tesla (T) as they are sometimes referred to. We can say then that one weber of magnetic flux spread evenly throughout a csa of one square metre results in a flux density of one weber per square metre, or one tesla. From this we can derive the following formula:

$$\text{Magnetic flux density } (B) \;=\; \frac{\text{Magnetic flux } (\Phi)}{\text{Area over which flux passes } (A)}$$

or algebraically:

$$B = \frac{\Phi}{A}$$

Example 10.1

The poles of an electric motor each have a cross-sectional area of 0.30 m². Calculate the flux density in the poles when the total flux per pole is 0.33 Wb.

$$B = \frac{\Phi}{A}$$

$$B = \frac{0.33}{0.30}$$

$$B = 1.1 \text{ tesla}$$

10.4 Electromagnetism

We have seen that non-magnetic materials such as copper or aluminium do not under normal conditions have a magnetic field surrounding them. However, if we allow an electric current to pass through them a magnetic field is created around them. The lines of magnetic force will behave in much the same way as those surrounding the permanent magnet.

The patterns of the magnetic fields produced are shown on Information Sheet No. 10B. Drawing (a) shows a section of a conductor with the current going *into the page* and this is depicted by a plus sign which is like the flight of an arrow going into the page. Here the magnetic field is travelling round the conductor in a clockwise direction. In drawing (b) the current is flowing *out of the page* and this is depicted by a dot which is like the point of an arrow coming out of the page. Here the magnetic field is travelling round the conductor in an anticlockwise

Information Sheet No. 10B Magnetic fields round conductors.

(a) Current going into the page.

(b) Current coming out of the page.

(c) Magnetic field due to a single loop.

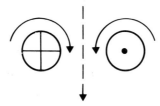

direction. A method of remembering the direction of the magnetic fields is Maxwell's corkscrew rule, where, if the corkscrew handle is turned clockwise representing the direction of the magnetic field, then the direction of the current will be in the direction in which the screw travels, i.e. into the page. Alternatively, if the corkscrew handle is turned anticlockwise the direction of the screw will be out of the page so the direction of the current flowing in the conductor will be out of the page too.

Drawing (c) of Information Sheet No. 10B shows two parallel conductors with the currents flowing in the opposite directions, as might be obtained by the use of a single loop of conductor. Here the current is entering the loop, circulating round the loop and coming out again. The lines of magnetic force produced do not cross and the resultant magnetic field is at right angles to the direction of the current.

Electromagnets

If we were to continue winding the conductor into the shape of a coil or solenoid a much stronger magnetic field would be produced, as shown in drawing (a) of Information Sheet No. 10C. The polarity of a solenoid depends on the direction in which the current is flowing and a method of remembering the direction of the magnetic field in is the *right hand grip* rule. If a conductor or coil is gripped in the right hand with the fingers pointing in the direction of the current, then the thumb will be pointing in the direction of the north pole, as shown in drawing (b) of Information Sheet No. 10C.

The field set up by a solenoid is a result of a current flowing through the turns of the coil. If we were to increase the number of turns with the same current flowing the effect would be to increase the magnetic flux (until saturation is reached). If we multiply the number of turns of the solenoid by the current flowing through it, we obtain the magnetomotive force (mmf) of the solenoid. This is measured in ampere turns (At or NI), but since the number of turns is simply a multiplier, A is often accepted.

Example 10.2

Find the mmf of a solenoid that has (1) 500 turns with a current of 2 A flowing, and (2) 1000 turns with a current of 1 A flowing.

(a)
$$\text{magneto-motive force} = \text{amperes} \times \text{turns}$$
$$\text{mmf} = 2 \times 500$$
$$\text{mmf} = 100 \text{ At}$$

(b)
$$\text{mmf} = 1 \times 1000$$
$$\text{mmf} = 100 \text{ At}$$

Information Sheet No. 10C Electromagnetic fields.

(a) Electromagnet

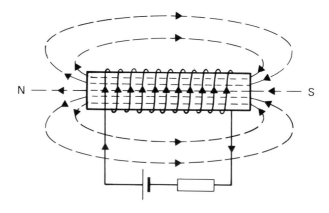

(b) The right hand grip rule

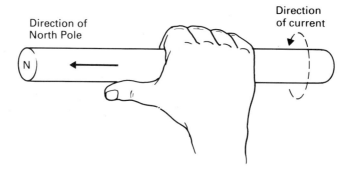

The coil mentioned above has only air in its centre and this offers a resistance to the flux path which we call *reluctance* (S). If we take a solenoid and slide a core of magnetic material into the centre of it, the magnetic flux (Φ) produced by the same current will be increased many times. This is because the magnetic material reduces the reluctance of the flux path and the magnetic flux produced is greater for the same number of ampere turns.

In simple terms we can say that *the magnetic flux of a circuit is proportional to the number of ampere turns and inversely proportional to the reluctance*, and this gives us the formula:

$$\text{Flux} = \frac{\text{Magnetomotive force}}{\text{Reluctance}}$$

Or algebraically

$$\Phi = \frac{At}{S}$$

If the above sounds vaguely familiar, it will come as no surprise to learn that many commentators draw direct comparisons with this and Ohm's law, often referring to it, indeed, as 'Magnetic Ohm's law'.

ELECTRIC

MAGNETIC

$$\text{Current} = \frac{\text{Electromotive-force}}{\text{Resistance}} \qquad \text{Flux} = \frac{\text{Magnetomotive force}}{\text{Reluctance}}$$

The design of magnetic circuits is complex and much of it is far above the level of Part 1 students; the topic is returned to again in Part 3. The above, however, will be enough to afford students a basic understanding of this part of the subject.

Applications for the electromagnet

We have seen that an electromagnet consists of a coil of insulated wire wrapped round an iron core. When an electric current is passed through the coil the core becomes strongly magnetised. Many applications require that once the current is switched off the magnetic effect should cease, for example lifting magnets, relays, electric bells, etc. Here the magnetic materials used are what we call *soft* magnetic materials, that is they are easily magnetised but lose their magnetism quickly. Examples of these soft magnetic materials are soft iron, silicon steel and stalloy, a steel alloy. Information Sheet No. 10D shows some of the uses of the electromagnet.

10.5 Force on a conductor in a magnetic field

We saw earlier in the chapter that a current-carrying conductor or coil will produce a magnetic field of its own. It is not surprising, therefore, that such a conductor or coil, when placed in a magnetic field, will be subject to a force. If a

Information Sheet No. 10D Applications for electromagnets.

1. Electromagnet for lifting scrap metal

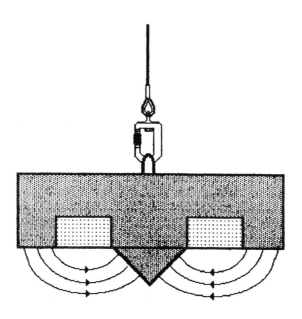

2. Trembler bell

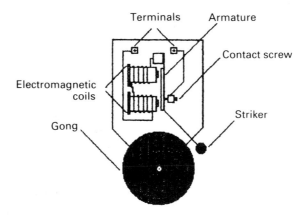

Information Sheet No. 10E Magnetic force on a conductor.

(a) Current going into the page.

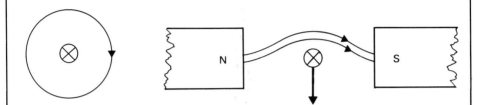

(b) Current coming out of the page.

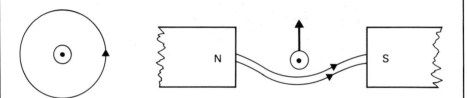

(c) Fleming's left-hand rule   Motion

Field

Current

current-carrying conductor lies in a magnetic field at right angles to the lines of magnetic force then a force will be exerted on that conductor. Diagram (a) on Information Sheet No. 10E shows such an arrangement. The direction of the force will depend on the direction of the magnetic field and the direction of the current through the conductor. In diagram (a) the current in the conductor is flowing away from the reader and is shown by the symbol + depicting the flight end of an arrow. The magnetic field around the conductor is in a clockwise direction and, in accordance with Fleming's left-hand rule, the conductor will be forced downwards. This is because the two magnetic fields complement each other above the conductor and are quite strong; however, below the conductor they tend to oppose each other, the magnetic field is weaker and the conductor is forced in a downwards direction.

If either the direction of the current or the direction of the magnetic field was to be reversed, then the conductor would be forced in an upwards direction (see diagram (b) of Information Sheet No. 10E where the direction of the current, depicted by a dot representing the point of an arrow, is flowing towards the reader). Diagram (c) on Information Sheet No. 10E depicts Fleming's left-hand rule; here we have three quantities mutually at right angles – the magnetic field direction, the current direction and the direction of motion. The rule is given below:

Fleming's left-hand rule for motors

First finger points in the direction of **F**ield
Se**C**ond finger points in direction of **C**urrent
Thu**M**b gives direction of **M**otion

We shall see later that this rule applies to electric motors as well, and in the UK we can remember that motors use the left-hand rule by remembering that *motors drive on the left.*

The strength of the force (*F*) acting upon the conductor depends upon three factors:

(a) the flux density (*B*) or strength of the magnetic field;
(b) the size of the current (*I*) flowing in the conductor; and
(c) the length (*l*) of the conductor in the magnetic field.

So that:

$$F = B \times I \times l \text{ newtons}$$

Where *F* is the force on the conductor in newtons, *B* is the flux density in teslas, *I* is the size of the current in the conductor in amperes and *l* is the length of the conductor in metres.

Example 10.3

A conductor 0.25 m long is placed at right angles to a magnetic field and is acted upon by a force of 10 N. If a current of 50 A is flowing what will the flux density be in teslas?

$$F = B \times I \times l \text{ N}$$
$$10 = B \times 0.25 \times 50$$
$$B = \frac{10}{0.25 \times 50}$$
$$B = 0.8 \ T$$

10.6 Principles of operation of a simple electric motor

The principle discussed above is, of course, the principle of operation of a simple electric motor. Diagram (a) on Information Sheet No. 10F shows an armature conductor placed in a magnetic field. For simplicity's sake it comprises a single conductor in the form of a loop, though in practice there would be many such loops. The magnetic fields around the conductors interact with the magnetic field of the permanent magnet as described above. Since one half of the loop has the current passing away from the reader and the other half of the loop has the current coming towards the reader, each half will be forced in opposite directions as shown and rotation of the loop is achieved.

In its simplest form an electric motor would consist of a coil rotating between two magnetic poles, as shown in Fig. 10.1. The current from a battery or d.c. supply, is fed to the coil via commutator and carbon brushes. This two-part commutator ensures that whichever side of the coil is nearest the north pole it will always have the current flowing in the same direction. This means that the force will always act in the same direction and rotation in the same direction will be maintained.

10.7 Induced emf in a moving conductor

We have seen from the above that the passing of a current through a conductor in a magnetic field results in movement of that conductor. If we reversed the process and moved the conductor in the magnetic field in such a way as to cut the magnetic flux it would be found that an emf had been induced in the conductor. Fig. 10.2 shows the magnetic field flowing from north to south and the conductor being moved sideways. The direction of the induced emf in the conductor is as indicated by the arrow.

The direction of the current can be determined by the use of Fleming's right-hand rule, as shown in Fig. 10.3. Here we have three qualities mutually at right angles – the magnetic field direction, the current direction and the direction of motion. The rule is given below:

Fleming's right-hand rule for generators

First finger points in the direction of Field
SeCond finger points in direction of Current
ThuMb gives direction of Motion

The magnitude of the induced emf (E) in the conductor depends upon:

(1) The flux density (B) of the magnetic field;

Information Sheet No. 10F Torque.

(a) The movement of a current-carrying loop placed in a magnetic field.

(b) The torque produced on a current-carrying loop in a magnetic field.

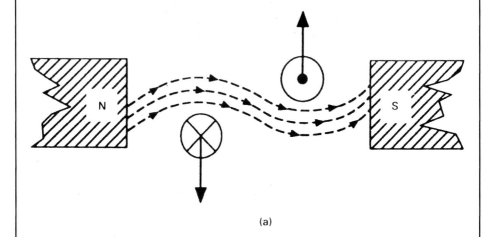

(a)

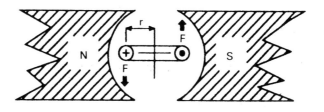

(b)

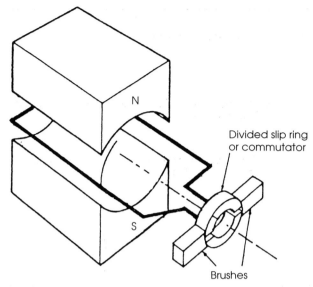

Fig. 10.1 Simple d.c. electric motor

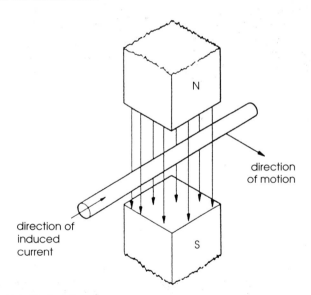

Fig. 10.2 Direction of induced emf

(2) The effective length of the conductor (l) at right angles to the magnetic field; and

(3) The velocity (v) at which the conductor cuts the magnetic flux.

So that:

$$E = B \times l \times v$$

Where B is measured in teslas, the effective length of the conductor l is measured in metres and the velocity v is measured in metres per second.

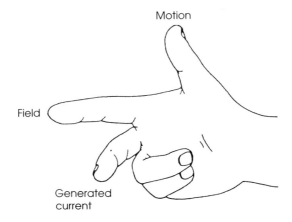

Fig. 10.3 Fleming's right-hand rule

Example 10.4

A conductor 0.5 m long is moved at right angles to a magnetic field of strength 1.1 T, at a velocity of 15 m/s. Find the value of the emf induced in the conductor.

$$E = B \times l \times v$$
$$E = 1.1 \times 0.5 \times 15$$
$$E = 8.25 \text{ V}$$

Example 10.5

The emf induced in a conductor of effective length 0.25 m, moving at a velocity of 10 m/s, is 2.0 V. Calculate the magnetic flux density in teslas.

$$E = B \times l \times v$$
$$2.0 = B \times 0.25 \times 10$$
$$B = \frac{2.0}{0.25 \times 10}$$
$$B = 0.8 \text{ T}$$

10.8 Principles of operation of a simple generator

Michael Faraday (1791–1867) observed that an emf is generated in a conductor which cuts across a magnetic field. If we arrange for conductors to be mounted on an armature and rotated in a magnetic field, then, where they cut the magnetic field, an emf will be generated in them. Information Sheet No. 10G shows a simple two-pole generator; only one armature conductor loop has been shown for simplification. As the loop is rotated by some form of mechanical energy or *prime*

Information Sheet No. 10G Simple a.c. generator (alternator).

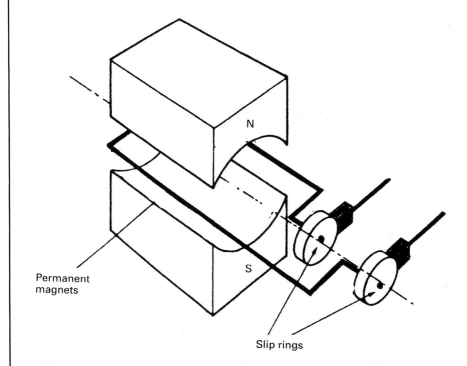

Information Sheet No. 10H The sinusoidal waveform.

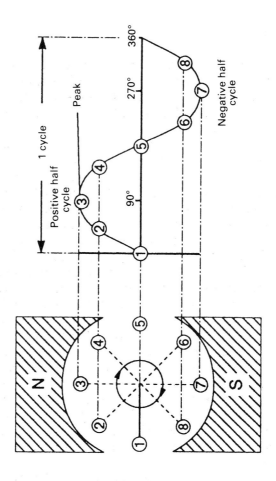

mover as it is called, each side of the loop passes the north pole and south pole in turn. An emf is induced in the loop and electrical energy is taken off by means of the two slip rings. The electrical energy generated is initially alternating (a.c.), as shown, and the machine is referred to as an alternator. However, if direct current (d.c.) is required then the slip rings can be substituted for by a two-part commutator and the machine is referred to as a dynamo. The commutator described has the effect of changing this alternating current (a.c.) into a direct current (d.c.) and can be regarded as a form of rectifier.

10.9 The sinusoidal waveform

When an emf is generated by an alternator it varies in both magnitude and direction according to the instantaneous position of the loop. If the strength of the field is uniform, then, in one revolution of 360 electrical degrees a waveform of the shape shown in the graph on Information Sheet No. 10H will be produced. The shape is that of a sine wave.

Information Sheet No. 10H shows a section of a simple alternator. When a conductor is in position 1 it is moving along parallel with the lines of magnetic flux, but it is not cutting them, and so no emf is induced in it. In position 2 the conductor is beginning to cut the magnetic flux at an angle and an emf is induced. At position 3 the conductor moves directly across the magnetic flux, cutting it at the maximum rate and the maximum emf is induced. Position 4 shows the conductor still cutting the flux but not at maximum so the induced emf will be less. At position 5 the conductor is moving along the lines of magnetic flux again and is not cutting them so no emf is induced. The emfs produced have been plotted on the adjacent graph to a base of time and it can be seen that in the positive half cycle the emfs have risen from zero to maximum and then fallen to zero again.

The coil continues to turn, taking the conductor towards the opposite pole of the magnet, and in position 6 we see the conductor again cutting the lines of flux at an angle so that an emf is generated, but this time in the negative half cycle. In position 7 the conductor is moving directly across the magnetic flux, cutting it at the maximum rate and a maximum emf is induced. Position 8 shows the conductor still cutting the flux but not at maximum so the induced emf will be less. At the final position in the cycle the conductor is moving along the lines of magnetic flux again and is not cutting them, so no emf is induced. From the graph we can see that in the negative half cycle the emfs have risen from zero to a maximum then fallen to zero once again.

When an alternator is rotated, the conductors (in which the emf is induced) are alternately swept across the north and south poles of the magnet. It will be seen from the graph on Information Sheet No. 10H that for each reversal of polarity, i.e. each half revolution of the coil, the direction of the induced emf reverses. Thus the current in the load changes direction twice per revolution of the coil. The reason for this can be explained by Fleming's right-hand rule. In the first half of the cycle the conductor moves through the magnetic field from left to right (positions 2, 3 and 4). In the second half of the cycle the conductor moves from right to left (positions 6, 7 and 8) so the induced emf will be in the opposite direction.

The time taken to complete one revolution or cycle is called the periodic time (symbol T). The number of cycles completed in one second is called the frequency (symbol f) and the unit in which this is measured is called the hertz (Hz). If there are f cycles in one second, each cycle will take $1/f$ second, so that:

$$T = \frac{1}{f} \text{ and } f = \frac{1}{T}$$

Example 10.6

Calculate the frequency of an a.c. system when the periodic time is 0.02 second.

$$f = \frac{1}{T}$$

$$f = \frac{1}{0.02}$$

$$f = 50 \text{ Hz}$$

10.10 Alternating quantities

The measurement of an alternating quantity can present some difficulties as the quantity concerned is continually changing both its magnitude and direction. There are a number of different ways of recording an alternating quantity and we will look at these.

Instantaneous value   This is the value of a waveform at a particular instant in time and will be different for different instants. We usually represent these with small symbols, i.e. v for voltage, i for current, etc.

Peak or Maximum value   The value of the highest point of the waveform, usually indicated by the symbols Vm for voltage, Im for current, etc.

Average or mean value   This is the average value of the waveform taken over one half cycle. If an average value was taken over a full cycle the positive and negative half cycles would cancel each other out. We usually represent these with the symbols V_{av} for voltage, I_{av} for current, etc.

Root mean square value   This is the value of a.c. current that produces the same amount of heat or does the same amount of work in the same time as that of an equivalent d.c. current. The root mean square (rms) value is sometimes referred to as the effective or virtual value and is indicated by the symbols V, I, etc.

Calculating alternating quantities

For practical purposes we can take the average or mean value to be 0.637 of the peak or maximum value and the rms value to be 0.707 of the peak or maximum value, as shown in Fig. 10.4.

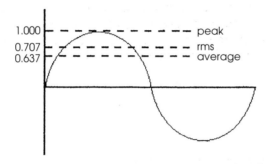

Fig. 10.4 Values of sine wave

The values for average and rms values can, however, be worked out by graphical methods. To find the average value the base line of the waveform is divided up into any number of equal parts. The example in Fig. 10.5 has been divided up into six parts for clarity, although the more parts that are used the greater the accuracy. At the centre of each part a mid-ordinate has been drawn from base line to curve. The average voltage in this case would be the average length of these lines.

$$V_{av} = \frac{V_1 + V_2 + V_3 \cdots + V_n}{n}$$

To find the rms value the procedure for dividing the waveform into parts and drawing mid-ordinates is the same as for finding the average value. Next square the value of each of the individual mid-ordinates. We then find the mean of the squared values by adding them and then dividing the total by the number of mid-ordinates. Finally we take the square root of the result and this gives us the root of the mean of the squared values.

$$V_{rms} = \frac{V_1^2 + V_2^2 + V_3^2 \cdots + V_n^2}{n}$$

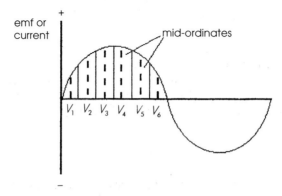

Fig. 10.5 The mid-ordinate method

Example 10.7

Find (a) the average value of voltage, and (b) the rms value of voltage of a waveform if the mid-ordinates are 30, 60, 90, 90, 60 and 30 V and the peak voltage is 94.19 V. Check your answers by using both methods of calculation.

(a)

Method 1

$$V_{av} = \frac{90 + 60 + 30 + 30 + 60 + 90}{6}$$

$$= \frac{360}{6}$$

$$= 60 \text{ V}$$

Method 2

$$V_{av} = \text{Max value} \times 0.637$$
$$= 94.19 \times 0.637$$
$$= 59.99 \text{ V}$$

(b)

Method 1

$$V_{rms} = \frac{90^2 + 60^2 + 30^2 + 30^2 + 60^2 + 90^2}{6}$$

$$= \frac{25\,200}{6}$$

$$= 4200 \text{ V}^2$$

$$= \sqrt{4200}$$

$$= 64.80 \text{ V}$$

Method 2

$$V_{rms} = \text{Max. value} \times 0.707$$
$$V_{av} = 94.19 \times 0.707$$
$$= 66.59 \text{ V}$$

The mid-ordinate method is not as accurate as using the constants although the more mid-ordinates used the greater the accuracy will be.

10.11 Inductance

We saw in Clause 10.8 above that when a conductor or coil is moved at right angles to a magnetic field so that it *cuts* the lines of magnetic force, an emf is induced into that conductor or coil. A similar result would be obtained if the magnet was moved and the conductor or coil was kept still, and this has been discussed in Chapter 9. An emf can, however, be induced without moving either the coil or the magnet.

If we place a coil carrying a d.c. supply in close proximity to a second coil in such a way that the second coil is linked with the flux emanating from the first coil, and then switch the current on and off, an emf will be produced in the second coil. If an instrument was connected to the second coil its pointer would be seen to swing over and then return to zero each time the current to the second coil was switched on and off, showing that an emf was being induced. The reason for this is that there has been a change in the flux linking the second coil, which has an emf induced in it just as if the coil had been moved in a magnetic field.

When a varying current such as an alternating current travels through a coil it sets up a varying magnetic field. This also has the effect of inducing an emf into a second coil that is linking with the flux. In other words as the field builds up and then collapses due to the varying current, the second coil is being *cut* by the field and an emf is induced into it. An emf will only be produced while the flux is changing and will stop once the flux becomes steady. This is the principle of operation of the transformer and is discussed more fully in Book 2 of this series.

If the flux linking a coil of N turns is Φ_1 (Wb) at an instant in time t_1 and Φ_2 at another instant in time t_2, the emf induced is:

$$\text{emf} = N \times \frac{\text{change in flux}}{\text{time required for change in flux}}$$

$$= N \times \frac{\Phi_2 - \Phi_1}{t_2 - t_1}$$

Example 10.8

The flux linking a coil of 24 turns changes from 0.05 Wb to 0.08 Wb in 0.003 seconds. Find the induced emf.

$$\text{emf} = N \times \frac{\Phi_2 - \Phi_1}{t_2 - t_1}$$

$$= 24 \times \frac{0.08 - 0.05}{0.003}$$

$$= 240 \text{ V}$$

10.12 Self-inductance

The magnetic flux linked with a coil may well be due to an electric current flowing in the coil itself. If there is a change of current in an individual coil there must be a change in magnetic flux through that individual coil and so an emf will be induced in the coil. This emf only exists while the current is changing and is called the emf of *self-induction*.

The direction of the induced emf is in accordance with Lenz's law, *such as to oppose the change in current producing the emf*. In other words if the current tends to rise the induced emf acts against the direction of current (back emf), while if the current tends to fall the induced emf acts in the same direction as the current. This is the principle of operation of the Choke in a fluorescent circuit and is discussed more fully in Book 2 of this series.

The symbol for inductance is L and the unit of inductance is the henry (H). A coil is said to possess *an inductance of 1 henry when current changing at the rate of 1 ampere per second (A/s) induces an emf of 1 V*.

If the self-inductance of a magnetic system is L henrys and the current changes from I_1 at an instant in time t_1 to I_2 at another instant in time t_2, the emf induced is:

$$\text{emf} = -L \times \frac{\text{rate of change in current}}{\text{time required for change in current}}$$

$$= -L \times \frac{I_2 - I_1}{t_2 - t_1}$$

It should be noted that the minus sign in the above equation indicates that the self-induced emf opposes the increase in current.

Example 10.9

A coil has a self-inductance of 20 H and a current through it changes from 1.20 A to zero in 0.01 seconds. Find the emf induced in the coil.

$$\text{emf} = -L \times \frac{I_2 - I_1}{t_2 - t_1}$$

$$= -20 \times \frac{0.00 - 1.20}{0.01}$$

$$= + 2400 \text{ V}$$

The + sign indicates that the emf was acting in the same direction as the applied emf, i.e. to prevent the reduction of the current. The above example helps to illustrate the very high emf that may be momentarily introduced when current in a highly inductive circuit is suddenly interrupted.

Energy stored in a magnetic field

We have seen that when the current in a highly inductive circuit is interrupted the current at that moment tends to be maintained. The explanation for this in simple terms is that the energy is being *stored* in the magnetic field while the current is increasing.

The derivation of the formula for energy stored is beyond the scope of the Part 1 student and the formula is given as:

$$\text{Energy stored} = 1/2 \times L \times I^2 \text{ J}$$

Example 10.10

A coil has an inductance of 0.4 H when carrying a current of 100 A. Find the energy stored in joules.

$$\text{Energy stored} = 1/2 \times L \times I^2 \text{ J}$$
$$= 1/2 \times 0.4 \times 10\,000$$
$$= 2000 \text{ J}$$

10.13 Mutual inductance

As we have seen from the above a change in current through a conductor or coil will not only induce an emf in the conductor or coil in the form of self-inductance, but also in any other conductor or coil linked by the changing flux. This effect is called *mutual inductance* (symbol M) and its unit is the same as that for self-inductance, namely the henry.

Given definition of mutual inductance   Two coils have a mutual inductance of 1 H if a rate of change of current of 1 A/s in one coil induces an emf of 1 V in the other coil.

So that if two coils A and B have mutual inductance of M henrys, the emf in coil A due to change in current in coil B can be found from:

$$\text{emf in coil A} = M \times \frac{\text{rate of change in current in coil B}}{\text{time required for rate of change}}$$

$$= M \times \frac{I_2 - I_1}{t_2 - t_1}$$

Example 10.11

Two coils have a mutual inductance of 6 H. If the current through one coil changes from 0.1 A to 0.6 A in 0.15 seconds, calculate the magnitude of the induced emf in the other coil.

$$\text{emf in coil A} = M \times \frac{I_2 - I_1}{t_2 - t_1}$$

$$= 6 \times \frac{0.6 - 0.1}{0.15}$$

$$= 6 \times \frac{0.5}{0.15}$$

$$= 20 \text{ V}$$

We shall return to the topic of mutual inductance when we discuss transformers in Book 2 of the series.

Test 10

Choose which of the four answers is the correct one.

(1) The unit of magnetic flux is the:

(a) Tesla;
(b) Weber;
(c) Wb/m²;
(d) mmf.

(2) The force on a conductor in a magnetic field is measured in:

(a) Newtons;
(b) Coulombs;
(c) Metres;
(d) Teslas.

(3) Magnetic flux density is measured in:

(a) Webers;
(b) mmf;
(c) Teslas;
(d) Coulombs.

(4) The emf induced in a conductor moved at right angles to a magnetic field is found from:

(a) $E = B \times 1 \times v$;
(b) $E = B \times I \times 1$;
(c) $E = V \times I \times R$;
(d) $E = I \times 1 \times v$.

(5) The waveform produced by an a.c. generator can be described as a:

(a) Saw tooth wave;
(b) Square wave;
(c) Pulse wave;
(d) Sine wave.

Chapter 11
The Heating Effect of an Electric Current

11.1 The effects of heat

When an electric current flows through a conductor or load it must overcome the resistance in the circuit and to accomplish this electrical energy is used and work is done. We have seen that energy cannot be destroyed, therefore the electrical energy used must be converted into some other form of energy and this shows itself as heat. To understand why this is so we must go back to our atomic theory with the electrons orbiting the nucleus. In addition to the orbital movement of the electrons, the orbits vibrate. When an electrical current flows through a conductor electrical energy is used and this increases the amount of vibration and the conductor becomes hotter than the surrounding air. If the current is increased above the current-carrying capacity of the conductor, then the conductor will become so hot that the insulation surrounding it will be irretrievably damaged.

When a substance is heated by passing an electric current through it, or applying some other form of heat, the following things can occur:

Change of dimension   There is an increase in atomic vibration (see above) and the substance expands (vibrating atoms require more space). Metals, liquids and gases expand when heated and contract when cooled.

Change in composition   If some substances are subjected to heat their whole composition is changed. For example, when a piece of sugar is heated, it first turns into a brown caramel and then into black carbon. We cannot reverse the process by cooling down the carbon.

Change of state   Atoms which vibrate strongly can break the mutual bonds in solids so that a solid becomes liquid and if the vibrations become even stronger can result in a liquid becoming a gas. For example, when ice is heated it becomes water and with further heating the water turns to steam.

Change in resistance   The resistance of metals generally increases as the temperature rises. They are described as having a positive temperature coefficient. Non-metals such as carbon, electrolytes and oxides of manganese experience a drop in resistance on temperature rise. They are described as having a negative temperature coefficient. Certain alloys such as Eureka, Nichrome and constantan have virtually constant resistance at all temperatures.

Heat transference   Heat is always transferred from a high temperature to a lower temperature; a current-carrying conductor which is getting hot will give heat out to its surroundings. The rate at which heat is transferred by a substance to its surroundings is determined by the difference in temperature between the two. A conductor installed in an environment having a high ambient temperature will not cool readily.

Thermo-electric effect   We saw in Chapter 10 how, in a thermocouple, two dissimilar metals joined together at one end and heated at the joint, produce a small electrical current.

11.2 Units of heat

The basic unit of work or energy in the SI system of units is the joule and, as heat is a form of energy it follows that quantities of heat are measured in joules. The amount of heat in joules required to raise the temperature of 1 kg mass of substance by 1°C is called the *specific heat capacity* (symbol *c*) of the substance.

Different substances have different specific heat capacities and Table 11.1 shows the specific heat capacities of some of the more common substances. Values of alloys such as brass or steel can vary according to their composition.

Table 11.1 Specific heat capacities of some common substances

Material	Specific heat capacities (J/kg°C)
Copper	390
Aluminium	946
Steel	450
Brass	380
Air	1010
Water	4190
Oil (transformer)	2140

Heat energy (J) = mass (kg) × specific heat capacity (J/kg°C) × temperature change °C
Or as an algebraic expression:

$$W = m\,c\,\theta \text{ joules}$$

Where:
W = heat required in joules
m = mass in kilograms
c = specific heat of substance
θ = temperature change in °C

Multiples and sub-multiples are used in conjunction with the joule as follows:

10^6 joules = 1 megajoule (MJ)
10^3 joules = 1 kilojoule (kJ)
10^{-3} joules = 1 millijoule (mJ)
10^{-6} joules = 1 microjoule (μJ)

Example 11.1

The temperature of 20 kg of copper rose from 30°C to 330°C in 5 minutes. Find (a) the quantity of heat in joules required to do this, and (b) the electrical power required.

(a)

$$W = m \, c \, \theta \, J$$

$$W = 20 \times 390 \times (330 - 30)$$

$$W = 2\,340\,000 \, J$$

(b)

The basic unit for energy is the joule and we have seen in Chapter 9 that one joule = one watt for one second. So that:

$$W = \text{watts } (P) \times \text{time in seconds } (t)$$

or

$$W = Pt$$

Transposing the formula for P:

$$P = \frac{W}{t}$$

$$P = \frac{2\,340\,000 \, J}{5 \times 60}$$

$$P = 7800 \, W \text{ or } 7.8 \, kW$$

11.3 Temperature

Temperature is not the same thing as heat, it is simply an indication of the level of heat or how hot a substance is. The SI unit of temperature measurement is the Kelvin (K). The Kelvin temperature scale has its zero point at *absolute zero*, which is, theoretically, the lowest temperature that can be obtained. This temperature scale is not very convenient to use so temperature is normally measured by a thermometer calibrated in degrees Celsius (°C) or centigrade as it is sometimes referred to. Kelvin and Celsius scales have the same spacing for temperature intervals and differ only in the position of the zero mark. The freezing point 0°C

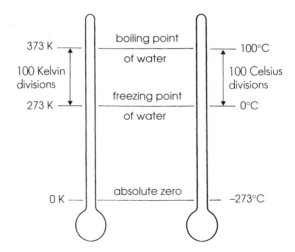

Fig. 11.1 Kelvin and Celsius scales

is equal to 237 K thus to convert degrees Celsius to the corresponding value in K it is only necessary to add 273. Fig. 11.1 compares the Kelvin and Celsius scales.

A few thermometers and thermostats are still marked in degrees Fahrenheit (°F). Fig. 11.2 compares the two scales, while the method of converting from one to the other is as follows:

$$°F = \left(\frac{9}{5} \times °C\right) + 32$$

$$°C = (°F - 32)\frac{5}{9}$$

Example 11.2

Convert the following:

(a) 15°C to °F and (b) 70°F to °C.

$$°F = \left(\frac{9}{5} \times 15°C\right) + 32 = 59°F$$

$$°C = (70°F - 32)\frac{5}{9} = 21°C$$

11.4 The thermometer

Any of the effects of heat described above which vary in a regular way with the rise or fall of temperature can be used to construct a thermometer or temperature-sensing device. Below are some of the more popular applications.

Mercury type   Mercury is a metal which has the unusual property of being in the form of a liquid at normal temperatures. This type of thermometer usually consists of a length of glass rod which has a fine bore down the centre of it. The lower end is formed into a bulb and this and part of the stem is filled with mercury as shown in Fig. 11.2. When the thermometer is heated the mercury expands and travels up the tube indicating the temperature which has been reached on the scale marked on the glass.

Alcohol type   Similar to the mercury type, it consists once again of a length of glass rod which has a fine bore down the centre of it. The lower end is formed into a bulb and this and part of the stem is filled with red-coloured alcohol. The volume of the liquid increases with a rise in temperature and this is indicated on the scale on the side of the glass tube. It provides a slightly different temperature range from that of mercury, being more suitable for the lower temperature range.

Metallic resistance type   A metallic resistance element is connected in series with a source of emf and a milli-ammeter. As the temperature of the element rises or falls, its resistance rises and falls accordingly. The milli-ammeter indicates the change in current and is calibrated in degrees so that the change in temperature can be read off directly.

Thermistor   These are temperature sensitive carbon resistors, whose resistance falls off rapidly when the temperature rises. The thermistor has a small current passing through it and any changes in its resistance are measured by an ohmmeter calibrated in degrees to that the temperature changes can be ready off directly.

Thermocouple   We saw in Chapter 10 that when two dissimilar metals are fused together at one end (which we call the hot junction) and wires are connected to the other ends which are not fused together (called the cold junction), any

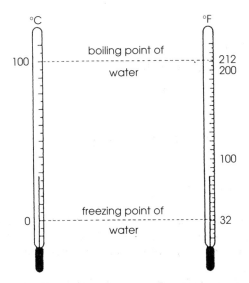

Fig. 11.2 Celsius and Fahrenheit scales

Information Sheet No. 11A Methods of sensing temperature change.

1. Electric iron thermostat.

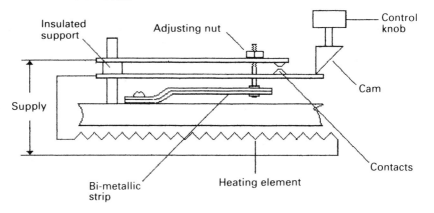

2. Oven thermostat.

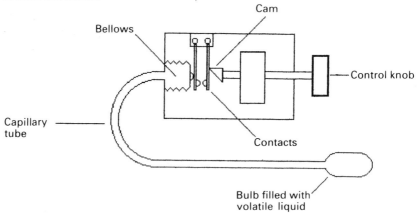

3. Thermistors.

 Bead-type, has a negative temperature coefficient.

 Disc-type, has a positive temperature coefficient.

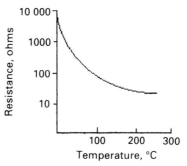

Resistance/temperature characteristics of a typical thermistor.

difference in temperature between the hot and cold junctions produces a difference of potential. Variations in temperature produce approximately proportional variations in voltage which can be used to give an indication of the temperature.

Room thermostats   These work on the principle that different metals have different expansion rates (see linear expansion below). If we bond together two dissimilar metals and make what we call a *bi-metal strip*, then subject the strip to a temperature change, the metals will expand or contract at different rates causing the strip to bend. The bi-metal strip is arranged in circuit to operate a set of contacts which will be either the *normally closed* type, which will open when the temperature rises, or the *normally open* type, which will close on temperature rise. A dial marked off in degrees operates a cam which either brings the contacts closer together, making them operate earlier, or moves them apart, which will delay the action.

Oven thermostats   These rely on the evaporation of a volatile liquid contained in a sealed bulb located in the oven. As the temperature rises the liquid evaporates, so raising the pressure inside the bulb. This pressure is transmitted via a capillary tube to a bellows in the thermostat's head; this in turn expands and operates a switch. As with the room thermostat a cam gives a variable temperature range.

Some of the ways that we can indicate the level of heat are shown on Information Sheet No. 11A.

11.5 The expansion of solids, liquids and gases

A close look at the above ways of recording temperature change will show that they use the different effects of heat to do this. Several use the changes in values of the electrical properties of components/devices and these are more fully discussed in *Electronics for Electrical Installations*, which accompanies this series. In general terms the other effects fall into four main categories:

(1) Linear expansion;
(2) Area expansion;
(3) Volume expansion;
(4) Expansion of gases.

If the original dimensions, the rise in temperature and the expansion of the substance, are measured, it will be seen that they are related to one another. The amount by which a unit length, area or volume of a body changes per degree change in temperature is called the coefficient of expansion of the substance.

If we take the *initial* length, area or volume of a body to be l_i, a_i and V_i respectively and the *final* length, area or volume of the body to be l_f, a_f and V_f after a change of t degrees in temperature, then the following clauses apply:

The coefficient of linear expansion   This is represented by the greek letter α; it is the amount by which unit length (1 m, etc.) of a material changes due to a one degree change in temperature. This can be represented algebraically as:

$l_f = l_i(1 + \alpha\, t)$

Where the increase in length $= l_i\alpha\, t$

The coefficient of area expansion   This is represented by the greek letter β; it is the amount by which unit area (1 m², etc.) of a material changes due to one degree change in temperature. This can be represented algebraically as:

$$a_f = a_i(1 + \beta\, t) \quad \text{where the increase in area} = a_i\beta\, t$$

The coefficient of volume expansion   This is represented by the greek letter γ; it is the amount by which unit volume (1 m³, etc.) of a material changes due to one degree change in temperature. This can be represented algebraically as:

$$V_f = V_i(1 + \gamma\, t) \quad \text{where the increase in volume} = V_i\gamma\, t$$

Gas   In this case, V_i is the volume at 0°C, t is the temperature above 0°C and the pressure is assumed to remain constant.

For solids: $\beta = 2\alpha$ approximately
and $\gamma = 3\alpha$ approximately

For a gas: $\gamma = \left(\frac{1}{273}\right)/°C$

Linear expansion is important to electrical engineers for many reasons and the syllabus of the City and Guilds asks that this is looked at in greater detail. Different substances have different coefficients of linear expansion and Table 11.2 shows the coefficient of linear expansion of some of the more common substances. Values of alloys such as brass or steel can vary according to their composition.

Table 11.2 Coefficients of Linear Expansion of some of the more common substances

Material	Coefficients of Linear Expansion (α)	
Copper	0.000017/°C	$(17 \times 10^{-6}/°C)$
Aluminium	0.000023/°C	$(23 \times 10^{-6}/°C)$
Steel	0.000011/°C	$(11 \times 10^{-6}/°C)$
Brass	0.000020/°C	$(20 \times 10^{-6}/°C)$

We have seen above that the change in length = original length × coefficient of linear expansion × temperature difference; some examples are provided below:

Example 11.3

A copper bus-bar has a coefficient of linear expansion of 0.000017/°C an.
measures 6 m long at 20°C. What would be the increase in length if its
temperature rose to 70°C?

The change in length $= l_i\, \alpha\, t$

$$= 6 \times 0.000017 \times (70\text{–}20)$$

$$= 0.0051 \text{ m or } 5.10 \text{ mm}$$

It can be seen quite clearly from this why the manufacturers of copper bus-bar
trunking have to insert expansion pieces in long runs of copper bus-bar.

Example 11.4

A gas boiler is controlled by a copper rod 200 mm long at 20°C, which operates
a switch when the temperature reaches 80°C. What will be the length of the rod
at 80°C?

$$l_f = l_i(1 + \alpha\, t)$$

$$l_f = 200(1 + 0.000017 \times 60)$$

$$l_f = 200.204 \text{ mm}$$

Example 11.5

An aluminium bus-bar is 8 m long at a temperature of 21°C. How long will it be
at 0°C?

The change in length $= l_i\, \alpha\, t$

$$= 8 \times 0.000023 \times (21\text{–}0)$$

$$= 0.00384 \text{ m}$$

Length of bar at 0°C $= 8 - 0.00384$

$$= 7.996 \text{ m}$$

Test 11

Choose which of the four answers is the correct one.

(1) The resistance of copper subjected to heat will:

(a) Remain the same;
(b) Fluctuate up and down;
(c) Increase;
(d) Decrease.

(2) Freezing point on the Celsius scale is equal to what on the Kelvin scale?:

(a) 270 K;
(b) 237 K;
(c) 273 K;
(d) 207 K.

(1) Metals subjected to heat expand at different rates and this is known as the:

(a) Coefficient of expansion;
(b) Specific gravity;
(c) Specific heat;
(d) Coefficient of temperature.

(4) The linear expansion of a copper bus-bar could be found from:

(a) Change in length $= B \times l \times v$;
(b) Change in length $= B \times I \times l$;
(c) Change in length $= V \times I \times R$;
(d) Change in length $= l_i \times \alpha \times t$.

(5) A thermistor whose resistance drops on rise in temperature has a:

(a) Negative temperature coefficient;
(b) Positive temperature coefficient;
(c) High specific gravity;
(d) Low specific gravity.

Chapter 12
The Chemical Effect of an
Electric Current

12.1 Secondary cells

As we have seen in Chapter 9 a *primary cell* is a source of electrical energy. The process of turning chemical energy into electrical energy cannot be reversed in this type of cell; *secondary cells* on the other hand can be re-charged, for, unlike primary cellss, secondary cells are not a primary source of electricity supply. They merely store the energy, and need to be charged by passing an electric current through the cell in the opposite direction to that when it supplies current to the load. In this way electrical energy is transformed into chemical energy. The current output capability of secondary cells, when connected together in the form of a battery, is much greater than that of primary cells, often exceeding 100 A for example when starting a car.

Secondary Cells are rated in ampere-hours (Ah); this is a measure of the amount of electricity they can store and the time taken to discharge it. For example, a battery rated at 60 Ah would give a steady current of 6 A for ten hours before the voltage dropped below an acceptable level (say 1.85 V). The same battery would give a continuous current of 10 A for six hours.

12.2 Lead–acid cells

If two lead plates are submerged in dilute sulphuric acid and are connected to a d.c. supply, certain chemical effects will take place. The positively connected plate will turn brown while the negatively connected plate remains slate grey. The action taking place is that the water (H_2O) in the solution separates, the hydrogen being released at the negative plate, while the oxygen combines with the lead of the positive plate and forms a small quantity of lead peroxide on the plate's surface. If the cell is discharged the plates turn grey due to the formation of lead sulphate on both plates. Repeated charging and discharging will build up sufficient amounts of the chemicals so that when the cell is discharged the discharge is big enough to be of practical use. The potential difference between the plates rises to approximately 2.7 V and the plates are then said to be *formed*.

The above process takes a long time and so commercially produced cells are produced in such a way that the customer can use them immediately. This is achieved by making the negative plate in the form of a grid and a paste made up of lead oxide pressed into the grids. The efficiency of the cell is further increased by designing the positive plate so that a greater surface area comes into contact with the electrolyte. This is done in a number of ways such as corrugating the plate

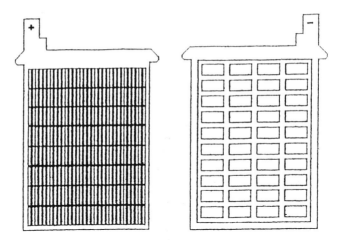

Fig. 12.1 Secondary cell grid-type plates

or making it into a fine grid. Fig. 12.1 shows examples of the lead–acid cell grids. The cell then requires only a preliminary charge before putting into service.

Care and maintenance of lead-acid cells

If these cells are regularly maintained there is no reason why they should not last for many years. The level of the electrolyte should never be allowed to fall below the level of the tops of the plates; some cells have marks on the case to indicate this level. If the electrolyte does fall, then it can be topped up using distilled water. Tap water should not be used as this often contains impurities that will shorten the life of your battery. If for some reason spillage takes place, this can be washed down with water containing soda. When mixing fresh electrolyte always add acid to the distilled water, not distilled water to the acid, otherwise there will be a violent reaction and someone could be injured. Rubber gloves should be worn during this operation, together with an acid-resistant apron or overall. If large quantities are involved eye protection of some kind should be employed. This could take the form of a protective screen or goggles.

When the lead–acid cell begins to discharge the electrolyte becomes weaker and its specific gravity drops. This can be measured by the use of a hydrometer of the type shown in Information Sheet 12A, which consists of a glass syringe containing a specially weighted float. The flexible nozzle of the hydrometer is placed into the cell, the bulb at the other end depressed and acid allowed to be drawn into the glass syringe. The float has a graduated scale and readings can be taken off; the higher the float is above the level of the electrolyte the higher will be the specific gravity. For a fully charged lead–acid cell the reading should be 1.28; this will drop to a level of 1.18 when discharged and is a good indication of the level of charge of a cell. The colour of the plates is another good indicator of the state of charge of this type of cell. In its fully charged state the positive plate is dark brown and the negative plate grey, these colours pale to light brown and pale grey as the cell

Information Sheet No. 12A Hydrometer.

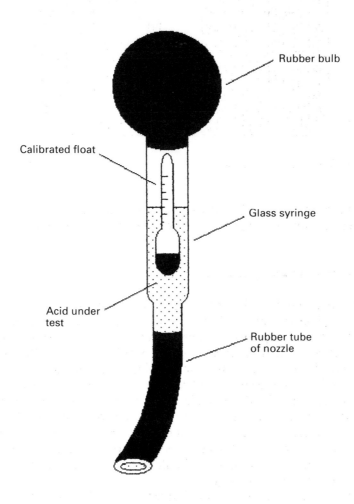

Rubber bulb

Calibrated float

Glass syringe

Acid under
test

Rubber tube
of nozzle

Hydrometer used to measure the
specific gravity of acid

discharges. Never allow a cell of this type to remain in a fully discharged state for any length of time, otherwise sulphation of the plates will take place. This coats the plates with a layer of sulphate, increasing the internal resistance, and reducing the capacity of the cell.

The battery should be kept clean and dry and all metalwork covered with a coat of petroleum jelly. Rooms that are used for batteries should be kept well ventilated and because of the gasses given off when charging is in progress, no smoking rules should be enforced.

12.3 Alkaline cells

There are two types of alkaline cell, *the nickel iron* and the *nickel cadmium*. In the nickel iron cell the positive plate is made of *nickel hydroxide* and the negative plate of *iron oxide*. In the nickel cadmium cell the positive plate is also nickel hydroxide; however, the negative plate is made of cadmium with a small amount of iron. The electrolyte for both types is potassium hydroxide. The plates for alkaline cells consist of flat nickel steel grids for the negative plates and either flat nickel steel grids or thin cylindrical tubes for the positive plates. These plates contain the active chemicals and are insulated from each other by ebonite rods; the whole assembly is placed in a welded steel container. As the construction of alkaline cells is for the most part steel, this gives the cells great mechanical strength; despite this they remain lighter than the lead–acid cells (see Fig. 12.2).

Recent developments have seen the emergence of small rechargeable nickel cadmium cells of the equivalent size to popular alkaline/zinc carbon non-rechargeable batteries. Some of these are capable of charging and discharging at high temperatures (maximum of 65°C) and where continuous overcharge on standby may be experienced. Typical applications include emergency lighting, alarm control panels and other emergency standby areas where high ambient temperatures are experienced. They are available in single 1.2 V single cells or multiple cell units. The cells should be constant current charged at a non-

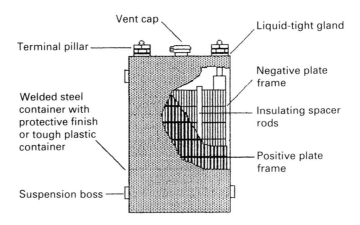

Fig. 12.2 Typical nickel–cadmium cell

continuous (cyclic) charge rate of 400 mA and a continuous (float) rate of 300 mA; nominal capacity is 4 Ah.

Care and maintenance of alkaline cells

Because of their robust construction alkaline cells require much less attention than lead–acid cells. Like other secondary cells the level of the electrolyte should be checked regularly and any losses made good by the addition of pure distilled water. Care should be taken when doing this that the utensils employed have not been used on lead–acid cells. The contamination of alkaline cells with acid of any sort will seriously affect their performance and may destroy them. If for some reason the electrolyte has to be renewed, then only the type specified by the manufacturer should be used as any slight difference in these can affect the performance of the cell. The cell should be discharged to a level recommended by the manufacturers and all the old electrolyte drained off before refilling commences. With this type of cell we do not have the problems with corrosive fumes and electrolyte splashes associated with other types of cell, though cleanliness is more important to avoid any chance of contamination of the electrolyte. All connections should be checked for tightness, and lightly greased with petroleum jelly. Under no circumstances should grease using animal or vegetable fat be used and no grease should be allowed to get into filler openings or ventilation holes.

The specific gravity of the alkaline cell does not go down during discharge and remains at a level of 1.2 during use. This means that the technique of checking the state of charge of the cell by use of the hydrometer cannot be used for these cells. It does have the advantage, however, that the cell has the ability to give practically its full rated capacity at high rates of discharge (see Fig. 12.3).

The electrolyte has no serious chemical action on any of the materials employed for the construction of the cell, or on any of the active materials used. This means that the cell can be left in a discharged state for long periods of time without this

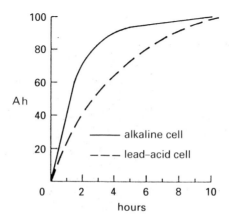

Fig. 12.3 Discharge rates of alkaline and lead–acid cells

having a detrimental effect on it. It also means that it has a very low self-discharge rate when left in an open circuit situation; this gives it a big advantage over the lead–acid cell.

12.4 Applications for secondary cells

Lead–acid cells   These cells are probably the most familiar type of secondary cell in use. Almost every motor vehicle on the road today uses one for its electrical system. Six of the cells are used together to form a 12-V battery. They are relatively inexpensive, can produce the high current necessary for the starting of a car, and have a high discharge voltage. They are not as robust as other types of secondary cell, however, and do require regular maintenance. They are often used for standby supplies, and emergency lighting and fire alarm systems.

Alkaline cells   More costly than lead–acid cells, nine of these cells are required to form a 12-V battery. Because they are very robust, can be charged and discharged quickly without damage, and left for long periods in a discharged state, they are often used in vehicles and equipment used by the armed forces. They are ideal for marine use, and where expense is not a major consideration they are being used more and more for standby supplies.

12.5 The charging of secondary cells

The charging of secondary cells is carried out by connecting them to some sort of regulated d.c. supply. This is usually a rectified a.c. supply, though it could be by means of d.c. mains, d.c. generator or a rotary converter. There are three commonly used methods of achieving this:

(1) *Constant current charging.* In this method the charge is begun at what is known as the starting current. This is continued until the voltage reaches a point some 20% above the cell's fully charged emf (exact figures should be obtained from the manufacturers). At this stage the cells will have begun to *gas*, so the current is reduced to a lower rate of charge and kept constant by varying the d.c. input until the completion of the charge (see Information Sheet No. 12B).
(2) *Constant voltage charging.* This method is more popular than the previous method as it can be automatically controlled. It consists of a constant voltage d.c. supply usually equivalent to the number of cells multiplied by 2.6, each of the cells having a separate resistance connected to them (full details should be obtained from the manufacturers as the values used are important). The charging current is high at the start of the charge, decreasing as the emf of the cells reaches that of the supply (see Information Sheet No. 12C).
(3) *Trickle charging.* In this system the battery of cells is maintained in a fully charged state by passing a very small charging current into the battery continuously. The battery is kept on charge continuously and the rate adjusted so that it is not being overcharged. In the event of discharge taking place, there

Information Sheet No. 12B Constant current charging.

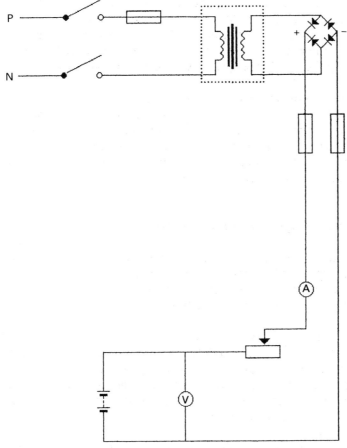

Circuit for the constant current method
of charging cells

Information Sheet No. 12C Constant voltage charging.

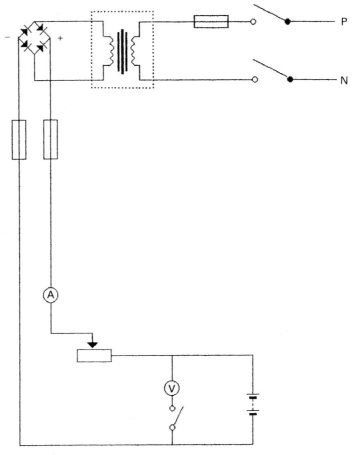

Circuit for the constant voltage
method of charging cells

is often a facility for rapid charging of the cells, so that they may be brought into service more quickly.

12.6 The connection of cells

When the cell is connected to an external circuit the emf falls to approximately 2 V; this is because of the effect of internal resistance. The current, when connected to an external circuit, flows through the resistance of the internal parts of the cell as well as through the resistance of the external circuit, and some of the cell's emf will be utilised to overcome this. The terminal voltage is the emf minus the voltage drop across the internal resistance. This can be determined by the use of the following formula:

$$V = E - Ir$$

Where E = the pd at the cell's terminals on open circuit
I = the current in the circuit when connected to the external load
r = the internal resistance of the cell
V = the pd at the cell's terminals when connected to the external circuit

Example 12.1

A cell of has a pd of 2.20 V on open circuit and 2.00 V when it is connected to an external circuit discharging at a rate of 10.00 A. Determine the internal resistance of the cell.

$$V = E - Ir$$
$$2.00 = 2.20 - (10.00 \times r)$$
$$2.00 + (10.00 \times r) = 2.20$$
$$10.00 \times r = 2.20 - 2.00$$
$$r = \frac{2.20 - 2.00}{10.00}$$
$$r = 0.02 \ \Omega$$

Example 12.2

A cell of internal resistance 0.02 Ω and pd 2.20 V is connected to an external circuit of resistance 0.98 Ω. What current will flow in the circuit?

$$V = E - Ir$$

It must be remembered that when the cell is connected to an external circuit the internal resistance of the battery (r) is added to that of the external circuit (R_e).

$$I = \frac{E}{r + R_e}$$

$$= \frac{2.20}{0.02 + 0.98}$$

$$= 2.2 \text{ A}$$

The connection of cells and batteries

Cells can be connected together in series to form batteries and a typical 12-V lead–acid battery would contain six cells. The internal resistances of the cells will, in effect, be in series, so if each had an internal resistance of 0.02 Ω then the total internal resistance of the battery (R_t) would be:

$$R_t = R_1 + R_2 + R_3 + R_4 + R_5 + R_6$$

$$R_t = 0.02 + 0.02 + 0.02 + 0.02 + 0.02 + 0.02$$

$$R_t = 0.12 \ \Omega$$

The individual pds would also be added and if the open circuit voltage of the individual cells was 2.20 V then the total pd on open circuit would be 13.20 V.

If the current provided by each cell is inadequate then identical cells can be connected in parallel. The pd on open circuit of a battery of cells connected in parallel would remain at the pd of an individual cell, i.e. 2.20 V, whereas the total internal resistance (R_t) would be found from:

$$\frac{1}{R_t} = \frac{1}{R_1} + \frac{1}{R_2} + \frac{1}{R_3} + \frac{1}{R_4} + \frac{1}{R_5} + \frac{1}{R_6}$$

$$\frac{1}{R_t} = \frac{1}{0.02} + \frac{1}{0.02} + \frac{1}{0.02} + \frac{1}{0.02} + \frac{1}{0.02} + \frac{1}{0.02}$$

$$\frac{1}{R_t} = \frac{6}{0.02}$$

$$R_t = 0.00333 \ \Omega$$

12.7 Electroplating

Electroplating is a practical use of the chemical effect of an electric current. A tank contains the plating solution, or electrolyte which usually consists of a solution of salt of the metal to be deposited. Mostly the anode consists of the plating metal, although in the case of chromium plating an insoluble anode is used, the chromium being provided by the electrolyte. The component to be plated is made the cathode. The whole unit is connected to a d.c. supply, as shown in Fig. 12.4.

The process is used for the protection from corrosion or decoration of metals. Zinc, tin, chromium and nickel are the most common metals used for plating. Low voltage supplies in the order of 4 to 20 V are used at a current density of 50 to 1000 A/m². A typical automatic nickel chrome plant operates at 9 V at 1000 A.

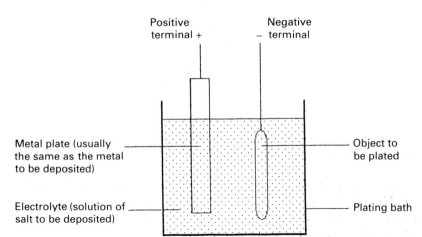

Positive
terminal +

Negative
– terminal

Metal plate (usually
the same as the metal
to be deposited)

Object to
be plated

Electrolyte (solution of
salt to be deposited)

Plating bath

Fig. 12.4 Electroplating

Test 12

Choose which of the four answers is the correct one.

(1) A primary source of electrical energy is a:

(a) Leclanché cell;
(b) Lead–acid cell;
(c) Nickel–cadmium cell;
(d) Nickel–iron cell.

(2) The electrolyte of a lead–acid cell contains:

(a) Citric acid;
(b) Acetic acid;
(c) Sulphuric acid;
(d) Nitric acid.

(3) Secondary cells are rated in:

(a) Ampere-hours;
(b) Volt-amps;
(c) Ampere-seconds;
(d) Watt-seconds.

(4) The specific gravity of a lead–acid cell can be measured by use of a:

(a) Graviometer;
(b) Voltmeter;
(c) Wattmeter;
(d) Hydrometer.

(5) On discharging the specific gravity of an alkaline cell:

(a) Drops;
(b) Remains the same;
(c) Rises;
(d) Fluctuates up and down.

Answers to the Tests

Test 1

(1) (b);   (2) (a);   (3) (c);   (4) (d);   (5) (a).

Test 2

(1) (c);   (2) (d);   (3) (b);   (4) (c);   (5) (d).

Test 3

(1) (d);   (2) (d);   (3) (a);   (4) (b);   (5) (c).

Test 4

(1) (d);   (2) (c);   (3) (a);   (4) (b);   (5) (d).

Test 5

(1) (c);   (2) (c);   (3) (a);   (4) (b);   (5) (d).

Test 6

(1) (b);   (2) (a);   (3) (b);   (4) (c);   (5) (d).

Test 7

(1) (a);   (2) (d);   (3) (b);   (4) (a);   (5) (b).

Test 8

(1) (b);   (2) (a);   (3) (d);   (4) (d);   (5) (a).

Test 9

(1) (b);   (2) (a);   (3) (d);   (4) (b);   (5) (a).

Test 10

(1) (b);   (2) (a);   (3) (c);   (4) (a);   (5) (d).

Test 11

(1) (c);   (2) (b);   (3) (a);   (4) (d);   (5) (a).

Test 12

(1) (a);   (2) (c);   (3) (a);   (4) (d);   (5) (b).

Index